Ahmed ISSOUFOU IMADAN

Regeneration of solar batteries

Ahmed ISSOUFOU IMADAN

Regeneration of solar batteries

Regeneration and desulphation processes and techniques for stationary solar lead-acid batteries

ScienciaScripts

Imprint

Cover image: www.ingimage.com

This book is a translation from the original published under ISBN 978-620-6-71362-3.

Publisher:
Sciencia Scripts
is a trademark of
Dodo Books Indian Ocean Ltd. and OmniScriptum S.R.L publishing group

120 High Road, East Finchley, London, N2 9ED, United Kingdom
Str. Armeneasca 28/1, office 1, Chisinau MD-2012, Republic of Moldova, Europe
Printed at: see last page
ISBN: 978-620-7-68195-2

To my dear mother, who has always prayed for me.

To my dear father for his advice.

SUMMARY

If the overall waste recovery system is not very advanced in African countries, the regeneration of used lead-acid batteries (BAPU) is even less developed. Large quantities of BAPU are exported every year to be recycled and reused in the manufacture of new batteries, which are very expensive in our countries and have a poor lifespan due to our climatic conditions.

Given the State of Benin's nationwide solar installation programs, the need for industrial, institutional and administrative back-up in the event of grid outages, and the growing number of off-grid solar installations in view of the country's considerable solar potential, the battery regeneration market is considerable. It's a deposit of stationary batteries that will be degraded every year.

Our work focused on the technical and economic feasibility of setting up a solar battery regeneration plant in Benin. After processing the data collected, we assessed a total potential of **15,000 12 V / 150 Ah gel batteries** and **5,000 OPzV 2V / 2,000 Ah gel batteries that could be** immediately exploited and regenerated.

With an operating life of **5 years** and a service cost of **35% compared with** the price of a new battery of the same type, the investment required for this project is **336,452,504** CFA francs, the Net Annual Value **475,426,769** CFA francs, the Internal Rate of Return **44%** with a Payback Period of **1 year 11 months 11 days** and a Profitability Index of **2.41**.

We can conclude that a project to install such a plant is both financially and technically feasible. Such a plant would not only help to protect the environment by massively reducing BAPU, it would also help on the social level by creating **twenty (20)** direct jobs, and on the economic level by generating good incomes. Its scope of action could be extended to the entire national territory, to reach the whole West African sub-region in the long term.

***Keywords**: Regeneration; Solar stationary batteries; Lead; BAPU; Benin; Solar PV street lamps; PV micropower plants.*

GENERAL INTRODUCTION

Solar batteries are the weak link in today's photovoltaic systems. They alone account for more than 30% of the initial investment cost. [1]. What's more, battery replacement is virtually the only operating cost for stand-alone PV systems.

On the other hand, the lifespan of batteries depends not only on their technology (open, sealed, Gel, AGM etc.) but also on operating conditions (temperature, deep discharge etc.). Properly sized and maintained, the best solar batteries have a lifespan ranging from 7 to 15 years [1]. Very often, lifetimes observed in the field are shorter.

In addition, several recent projects (PRODERE, PROVES, etc.) have enabled the installation of several thousand street lamps, micro-power stations and individual solar photovoltaic kits in several West African countries (Benin, Burkina Faso, Guinea, Niger, Togo, etc.). After less than 5 years in operation, many of these installations have broken down due to a lack of batteries. This means that large quantities of solar batteries have already been damaged and are being replaced for lack of alternatives.

An important question arises here: What is currently being done with damaged solar batteries at the end of their life?

The 1ère most common way is to sell these batteries at very low prices to craftsmen who break them to remove the lead and discharge the sulphuric acid and plastic components into the surrounding environment (sea, river, landfill etc. ......). The second way consists in collecting and transporting these batteries to recycling plants (these plants emit large quantities of greenhouse gases and lead which is dangerous to human health and the environment) where they are fragmented and treated separately for lead, plastic components and sulphuric acid, whereas there is a more environmentally-friendly, economical and efficient treatment method called regeneration, which gives batteries a second life so that they can be used again for the same functions.

It is therefore with a view to providing this economic and ecological solution that we are embarking on this technical and economic feasibility study for the establishment of a regeneration chain for stationary solar batteries, both locally (in Benin) and in the West African sub-region.

The study is structured into four chapters as follows:

- Chapter 1: Overview of batteries ;
- Chapter 2: presents the causes and failure modes of these batteries, as well as an assessment of the potential for damaged lead-acid batteries in Benin;
- Chapter 3: a technical and comparative study of different battery regeneration processes;
- Chapter 4: provides a cost-benefit analysis for the design of a battery regeneration unit in Benin.

The study ends with a general conclusion that gives a particular view of everything that has been studied.

CHAPTER 1: GENERAL INFORMATION ON BATTERIES

INTRODUCTION

The storage of electrical energy is an operation that involves placing a certain quantity of energy in a given location, to be available when it is needed. The techniques used to store electrical energy are numerous. These include [2] mechanical (energy transfer storage, compressed air), electrochemical (battery, hydrogen storage, methane), thermal (latent heat, sensible heat) and electrical (supercapacitor, capacitor) storage technologies.

To store electrical energy significantly and use it over long periods, it is sometimes necessary first to transform it into another intermediate and storable form of energy (potential, kinetic, chemical or thermal), as is the case with electrochemical accumulators, for example.

An electrochemical accumulator is an electrochemical generator capable of supplying electrical energy from chemically stored energy. This energy conversion is reversible for an accumulator, so it is rechargeable, unlike a battery. [3]. Several electrochemical couples characterize different battery technologies, including the lead-acid battery, the subject of this study.

This chapter provides an overview of all the key elements on which our study is based. It will be organized into five parts:

- The first part explains how batteries work;
- The second part presents the typology of batteries;
- The third part provides a general introduction to lead acid batteries;
- The fourth section describes the processes that can be used to recover end-of-life lead-acid batteries (BAPU);
- The fifth section describes the type of financial analysis and the main profitability evaluation formulas used in this work.

1.1 BATTERY OPERATING PRINCIPLE

A battery consists of a series and/or parallel assembly of electrochemical cells. Each accumulator consists of two electrodes, one positive and one negative, separated by an electrolyte. [4].

It works as follows: electrochemical reactions involving oxidation or reduction of the electrodes' active materials take place at the battery's electrode-electrolyte interfaces:

- At the anode (negative discharge electrode), an oxidation reaction takes place with redox potential □1 according to equation :

$$M1 \rightarrow \square 1^{n+} + \square\square\text{-} \tag{1.1}$$

- At the cathode (discharging positive electrode), electrons released at the anode pass through the external circuit to reach the cathode, where a reduction reaction of redox potential $E2$ takes place according to equation :

$$\square 2^{n+} + \square\square\text{-} \rightarrow \square 2 \tag{1.2}$$

$M1$ and $M2$ represent the active species at the anode and cathode respectively. The electrolyte transports the ionic species involved in the overall redox reaction, which is written as :

$$M1 + \square 2^{n+} \rightarrow \square 1^{n+} + \square 2 \tag{1.3}$$

Transport generates an electromotive force:

$$E = E2 \text{ - } \square 1 \tag{1.4}$$

During charging, the phenomenon is reversed. Figure 1.1 illustrates the general operating principle of an electrochemical accumulator.

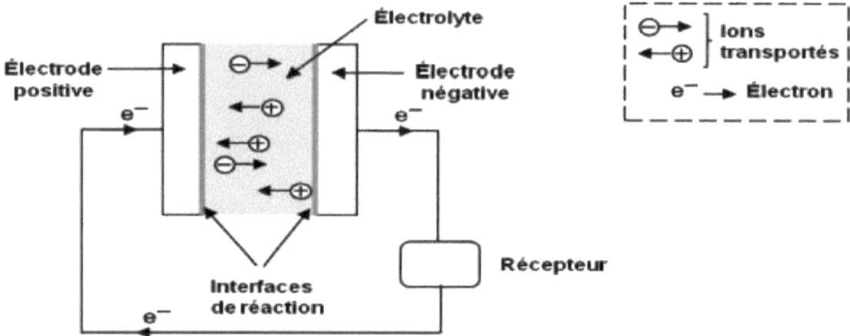

Figure 1.1: Operating principle of an electrochemical storage battery [4]

1.2 DIFFERENT TYPES OF ELECTROCHEMICAL BATTERIES

There are different types of electrochemical battery, each with its own specific operating profile, and consequently different technologies.

1.2.1. **Battery types by field of application**

Major applications include starter batteries and industrial batteries (stationary and traction batteries). [5] :

1.2.1.1. Starter batteries

They are designed to power a starting, lighting or ignition system. They are used in internal combustion engines for two- and four-wheel recreational vehicles, as well as for heavier vehicles such as trucks and tractors, generators, boats and aircraft. [5].

1.2.1.2. Industrial batteries

They are designed for exclusively industrial or professional use. A distinction is made between :

1.2.1.2.1. Stationary batteries

Stationary batteries are batteries designed to be used in a permanent location, unlike the batteries used in portable equipment (cell phones, laptops, digital cameras). [6]. They are used in back-up systems in the event of main grid failure (telecommunications, railroad signalling, hospitals, etc.) and in photovoltaic or wind power installations, as well as for power grid stabilization. [5].

1.2.1.2.2. Traction batteries

They are mainly used in equipment such as fork-lift trucks, handling equipment, wheelchairs, etc. [5]etc......

Table 1.1 below shows the most important characteristics for battery design and selection, depending on the application.

Table 1.1: Battery selection characteristics by application [5]

Battery type	Application examples	Important design and selection features	Technologies used or planned
Start	Automobiles, generators	Cost Maintenance-free power-to-weight ratio	Open acid lead thin plates
	Trucks, tractors	Cost Mass power Mass energy	Open acid lead
	Marine engines	Cost Mass power Mass energy	Sealed lead-acid Lithium-ion
Stationary	Inverters for network security Emergency lighting	Power-to-weight ratio Floating service life	Sealed lead Nickel-cadmium
	Storage for autonomous energy systems (solar, wind, power navigation)	Mass energy Maintenance-free cycling	Open lead Sealed lead Nickel-cadmium Lithium-ion
	Storage for grid-connected energy systems	Power density Mass energy Maintenance-free cycling	Sealed lead Sodium-sulfur Lithium-ion
Traction	Handling equipment, Wheelchairs Electric-assist bikes	Cost Mass energy	Open lead-acid tube sheets Open acid lead reinforced flat plates Recombination lead Lithium polymer

1.2.2. **Battery types by technology**

Batteries also differ in terms of their electrochemical pairs and some key characteristics specific to each technology, namely energy and power (mass or volume). Figure 1.2 shows the different types of battery, according to technology.

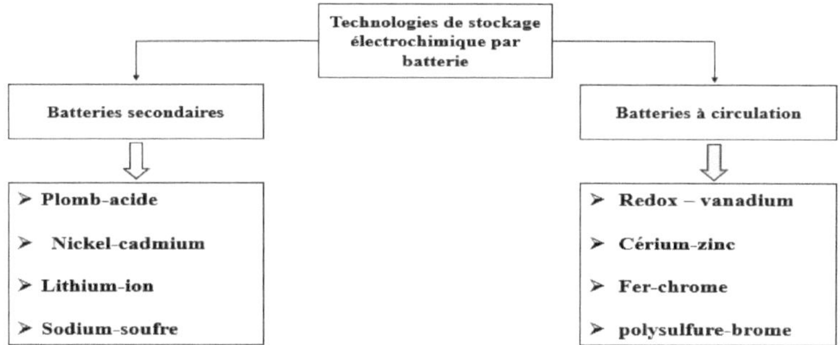

Figure 1.*2*: Different battery technologies [7]

Of all these technologies, the most widely used for stationary energy storage are : Lead-acid, nickel-cadmium and lithium-ion batteries [4].

*1.2.2.1.*Lead-acid battery

The lead-acid battery is the oldest and most widely used technology. Its electrolyte is an aqueous solution of sulfuric acid, the positive electrode consists of lead oxide PbO_2 and the negative electrode of lead Pb. The reaction involved is as follows:

$$Pb + PbO_2 + 2H_2SO_4 \rightleftarrows 2PbSO_4 + 2H_2O \quad (1.5)$$

The terminal voltage of a lead-acid battery varies from 1.7V (minimum state of charge) to 2.5V (maximum state of charge). [4].

*1.2.2.2.*Nickel cadmium battery

Nickel-cadmium batteries outperform lead batteries in terms of capacity and service life. However, their price is considerably higher and their voltage lower (1.15V to 1.45V) than that of lead batteries.

The electrolyte is potassium-based, the positive electrode nickel hydroxide and the negative electrode cadmium. These elements react as follows:

$$2NiO(\square\square) + Cd + 2H_2O \rightleftarrows 2Ni(\square\square)_2 + Cd(\square\square)_2 \qquad (1.6)$$

*1.2.2.3.*Lithium-ion (Li-ion) battery

During charging, lithium ions are inserted into the structure of the graphitized carbon negative electrode. During discharge, the anode releases these ions, which are deposited in the cathode structure. The overall electrochemical equation is as follows:

$$Li + MI \rightarrow LiMI \qquad (1.7)$$

MI the insertion material (graphite, coke, etc.) placed at the positive electrode. Compared with lead-acid batteries, lithium-ion batteries are maintenance-free. They have a long service life and are more resistant to external conditions. The voltage of a lithium-ion battery varies from 2.5V (minimum state of charge) to 3.7V (maximum state of charge) *[4]*.

The main comparisons of some battery technologies are presented in Table 1.2 below.

Table 1.2: Comparison of battery characteristics

Features	**Lead - acid**	**Nickel - cadmium**	**Lithium-ion**
Energy density (Wh/kg)	25 - 45	20 - 60	80 - 150
Power density (W/kg)	80 - 150	100 - 800	500 - 2000
Voltage range (V)	1,7 - 2,5	1,15 - 1,45	2,5 - 3,7
Discharge time	15mn - 100h	15mn - 100h	45mn - 100h
Storage time	> 1 month	< 1 month	Several months
Yield (%)	60 - 98	60 - 80	90 - 100
Service life (number of cycles)	300 - 1500	300 - 1500	> 1500
Current application	Starting, pulling and stationary	Stationary high-power, high-temperature applications	Mobile, satellite and stationary phones.
Maturity level	Very good	Approved	Very good
Benefits	-Electrochemical system with the lowest costs;	Operation at higher temperatures	-Good depth of discharge, -Excellent efficiency

Features	Lead - acid	Nickel - cadmium	Lithium-ion
	-Recyclable materials (nearly 100% for lead)		
Disadvantages	-Service life highly dependent on operating conditions (temperature in particular); -Toxic materials (lead).	Value for money compared to two other winning technologies	-High costs ; -Lithium resource issues

1.3. GENERAL INFORMATION ON LEAD ACID TECHNOLOGY

1.3.1. **Composition of a lead-acid battery**

It should be noted that, at the end of this section, the stationary battery is the type of battery considered for the remainder of this study. In general, the typical structure of a BAP composition can be seen in the following figure 1.3.

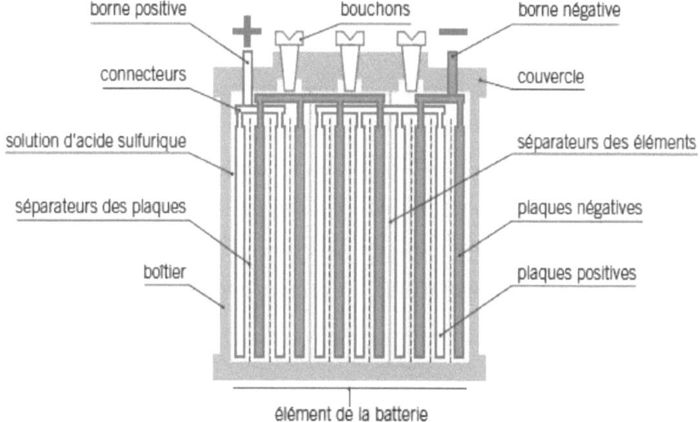

Figure 1.3: Typical structure of a lead-acid battery [8]

Although battery manufacturers employ different construction techniques and technologies, all BAPs include at least the five basic components [9] as follows:

- A resilient plastic **container**;

- Internal positive and negative lead **plates**;
- Porous plastic plate **separators**;
- **Electrolyte** (a dilute solution of sulfuric acid and water, also known as "battery acid");
- **Lead terminals** (the points of contact between the battery and the device it powers).

After these basic elements, there are also the microelements shown in figure 1.3 above. These microelements are detailed as follows [8], [9] :

- The positive and negative terminals, made of lead (in some applications, brass or copper may also be used), to which the external electricity consumer's devices are attached;
- Plugs (found on older batteries, one plug for each battery cell) for topping up with distilled/deionized water if necessary, and providing an escape route for the gases that form in the cells (new batteries are generally sealed);
- Lead connectors that establish electrical contact between plates of the same polarity, as well as between separate elements;
- The cover and casing, made of polypropylene or copolymer, contain the battery cells;
- A solution of sulphuric acid (the battery's electrolyte) ;
- Element separators, usually part of the housing and made of the same material, provide chemical and electrical insulation between the elements;
- Plate separators, made of PVC or another porous material, which prevent physical contact between two adjacent plates while allowing the ions in the electrolytic solution to move freely;
- Negative electrodes (plates): these are lead metal grids whose cells are filled with a lead dioxide paste ($PbO2$) ;
- Positive electrodes (plates): metallic lead plates coated with $PbO2$ paste;
- Battery cells, i.e. a series of positive and negative plates arranged in alternation and isolated from each other by plate separators.

1.3.2. **How a lead-acid battery works**

When a BAP supplies electrical energy to an external device, several chemical reactions take place simultaneously. A reducing reaction occurs at the positive electrodes (cathode) when lead peroxide (PbO_2) is converted to lead sulfate ($PbSO_4$). On the negative plates (anode), on the other hand, an oxidizing reaction takes place, converting metallic lead into lead sulfate. The electrolyte, sulfuric acid (H_2SO_4), provides the sulfate ions for both half-reactions and acts as a chemical bridge between them. For every electron produced at the anode, one is consumed at the cathode [10]. The reactions involved are as follows [10] :

- Anode :

$$Pb + SO_4^{2-} \rightarrow PbSO_4 + 2e^- \tag{1.8}$$

- Cathode :

$$PbO_2 + SO_4^{2-} + 4H^+ \rightarrow PbSO_4 + 2H\,O_2 \tag{1.9}$$

- Full reaction:

$$Pb + PbO_2 + 2H_2SO_4 \rightleftarrows 2PbSO_4 + 2H_2O \tag{1.10}$$

1.3.3. **State of discharge and charge of a lead-acid battery**

1.3.3.1. Discharge status

When the battery is in a state of discharge, the active materials of the positive electrode, lead dioxide, are transformed into lead sulfate. Similarly, the negative electrode, which consists of lead, will also be transformed into lead sulfate. The electrolyte, sulfuric acid, will be consumed by reacting with the active materials. The reaction equations for the positive electrode and the negative electrode in the state of discharge are demonstrated below. [11] :

- Cathode *:*

$$PbO_2 + 4H^+ + 2e^- \rightleftarrows Pb^{2+} + 2H\,O_2 \tag{1.11}$$

- Anode :

$$Pb \rightleftarrows Pb^{2+} + 2e^- \tag{1.12}$$

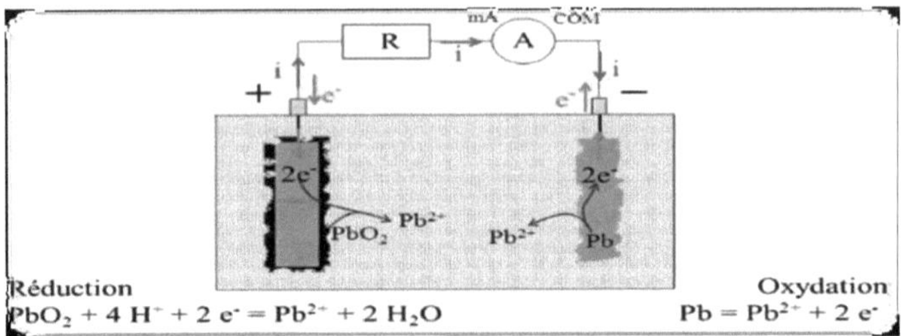

Figure *1.4*: Lead-acid battery state of discharge [10]

*1.3.3.2.*Load status

When the battery is in the state of charge, lead sulfate is transformed into active materials. The equations involved in this process are shown below [11] :

- Cathode :

$$PbSO_4 + 2H_2 0 \rightleftarrows PbO_2 + HSO_4^- + 3H^+ + 2e^- \quad (1.13)$$

- Anode :

$$PbSO_4 + H + \rightleftarrows Pb + HSO_4^- \quad (1.14)$$

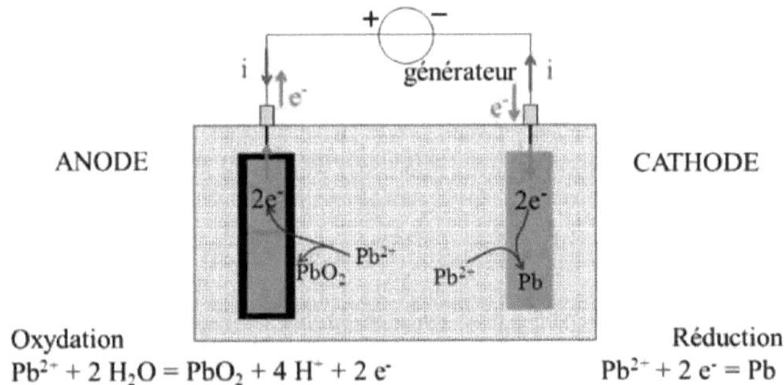

Figure *1.5*: Lead-acid battery state of charge [10]

Theoretically, a lead-acid battery can be used for a very long time if the chemical reaction equation shown in equation (1.10) can proceed completely. The chemical reaction taking place in lead-acid batteries has been shown to be somewhat reversible.

As a result, lead sulfate gradually crystallizes and deposits on the electrodes. This phenomenon is known as sulfation. It generally leads to a short service life for lead-acid batteries [11].

1.3.4. **Key features**

The main characteristics of a battery include: rated capacity, rate of discharge, rated voltage, state of charge, depth of discharge, cycle life, self-discharge, energy efficiency and internal resistance.

- **Rated capacity** [5], [12]

This is the total number of Ampere hours (Ah) available when the battery is discharged at a certain discharge current (specified as rate C/time, the time in hours) from 100% state of charge to cut-off voltage. Capacity is calculated by multiplying discharge current (in Amps) by discharge time (in hours), and decreases with increasing discharge rate.

- **Discharge regime** [13]

The discharge rate indicates the time in hours it takes to discharge the battery's nominal capacity. We generally speak of C_5 (battery capacity discharge time corresponds to 5 hours) for traction batteries and C_{24} for solar batteries (battery capacity is generally discharged in 24 hours).

- **Rated voltage** [5], [12]

This is the battery's signal or reference voltage, sometimes also referred to as the battery's "normal" voltage.

- **State Of Charge (SOC)** [5], [12]

It shows the battery's real-time capacity as a percentage of its rated capacity.

- **Deeph of discharge (DOD)** [5], [12]

This is the percentage of energy delivered by a battery in relation to its rated capacity. A discharge of more than 80% DOD is called a deep discharge.

NB: It is important to note that the sum of a battery's state of charge (SOC) and depth of discharge (DOD) is always 100%. [14]. Figure (1.6) below explains the evolution of SOC and DOD as a function of battery capacity.

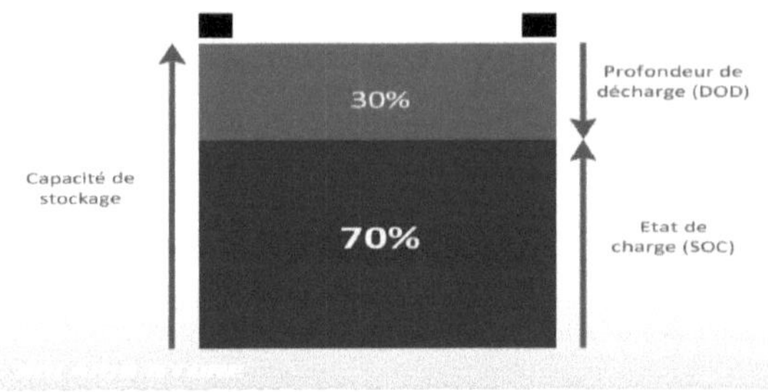

Figure *1.6:* State of charge and depth of discharge behavior as a function of battery capacity [14]

➢ **Cycle life** [12], [14]

Cycle life is the number of cycles a battery can withstand before deteriorating. It is expressed in number of charge/discharge cycles and is highly dependent on depth of discharge. Figure (1.7) below illustrates the cycling of a battery over 24 hours as a function of the state of charge of a solar array.

Etat de charge de la batterie – Battery SOC (State of charge)

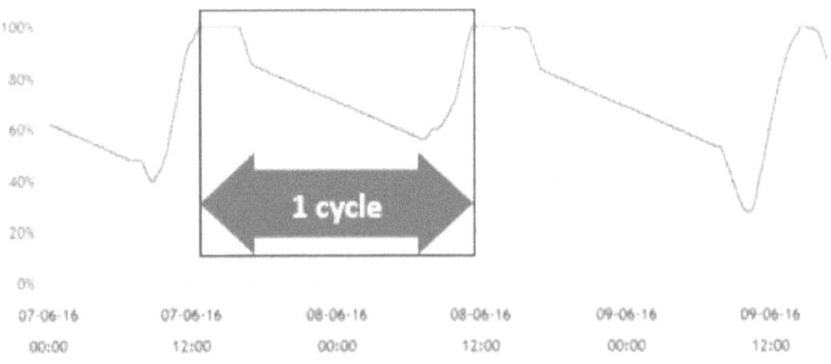

Figure *1.7*: Battery cycle [1]

- **Self-discharge** [12]

Self-discharge is the loss of battery capacity at rest. It is expressed in % per month and increases with age and temperature.

- **Energy efficiency** [12]

A battery's energy efficiency represents the amount of energy that can be used as a percentage of the amount of energy needed to store it.

- **Internal resistance** [11]

Internal resistance is a good indication of the state of charge. Measuring a battery's internal resistance can be used to monitor the main modes of battery failure - sulfation, grid corrosion and drying-out, etc......).

1.3.5. **Typology of lead-acid batteries**

Different lead-acid battery technologies are distinguished according to the type of electrodes (flat plates or tubular plates) and electrolyte (liquid or gelled). Referring to the nature of the electrolyte, a distinction is made between open batteries and closed or sealed batteries (Valve Regulated Lead Acid battery, VRLA battery). [10], [12], [15]. Figure (1.8) below shows the different types of lead-acid batteries.

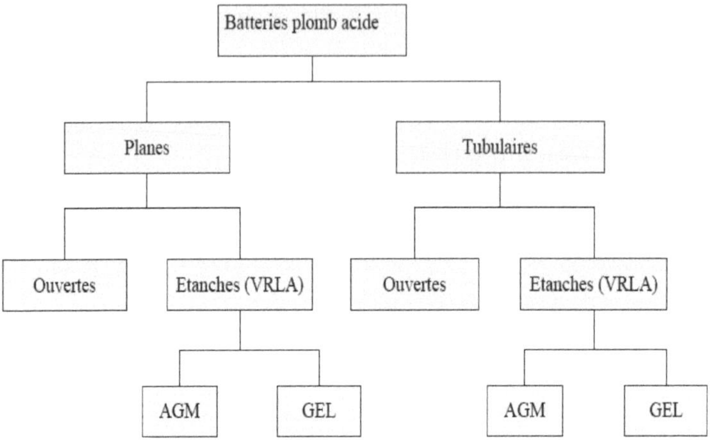

Figure *1.8*: Different types of lead-acid batteries [12]

*1.3.5.1.*Open lead-acid battery

During operation of an open lead-acid battery, the gases produced by the secondary reactions of water decomposition escape naturally through the holes in the plugs. The release of dihydrogen in the battery storage area is a source of danger, as its mixture with ambient air is potentially explosive above 4% by volume. [10].

For emergency stationary applications, installation in specific ventilated rooms is mandatory. The open batteries produced today (based on grids with high oxygen and hydrogen overvoltage) are often described as "maintenance-free" or "maintenance-free".

These names have been chosen because electrolyte consumption is so low that the original electrolyte reserve is sufficient to ensure the battery's correct operation throughout its life. [10].

*1.3.5.2.*Waterproof battery

Valve-regulated lead-acid batteries (VRLA), also known as "sealed lead-acid (SLA)", "gel cell" or "maintenance-free", are low-maintenance, rechargeable sealed lead-acid batteries. They limit the entry and exit of gas into and out of the cell, hence the term "valve-regulated" [15].

VRLA batteries are the most widely used because of their high power density and ease of use. VRLA batteries are considered "sealed" because they do not normally allow the addition or loss of liquid. The term VRLA derives from the use of safety valves to release pressure when a fault condition causes internal gas to build up faster than it can recombine [15].

There are three main types of VRLA batteries [15] :

1.3.5.2.1. Sealed wet cell with regulated valve

This type of battery contains acid in liquid form, in the same way as an open lead-acid battery, but the VRLA wet cell battery case is better sealed [15].

1.3.5.2.2. AGM batteries (AGM stands for Absorbed Glass Mat)

AGM batteries differ from open lead-acid batteries in that the electrolyte is retained in the glass mats, rather than freely flooding the plates. The fibers that make up the fine glass mat do not absorb and are not affected by the acid electrolyte. [15].

1.3.5.2.3. Gel batteries

Gel cells add silica dust to the electrolyte, forming a thick, putty-like gel. They are sometimes called "silicone batteries". Unlike an open lead-acid battery, these batteries do not need to be held in a vertical position. Gel batteries reduce electrolyte evaporation, spillage (and consequent corrosion problems) common to the open battery, and offer better resistance to shock and vibration [15].

As mentioned above, figure (1.9) shows the three types of VRLA batteries from left to right: sealed lead-acid battery; AGM battery; gel battery.

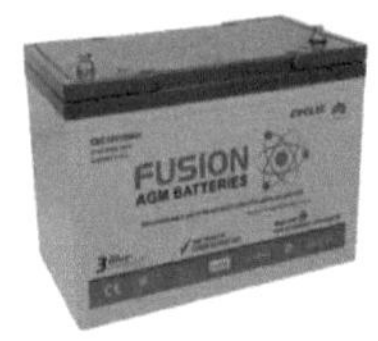

Figure 1.9: VRLA sealed, AGM and gel batteries [15]

Table 1.3 below shows the main comparisons between two major lead-acid battery technologies. [10].

Table 1.3: Comparison of two lead-acid battery technologies

Lead battery type	**Open**	**Waterproof**	
Electrolyte	Liquid	Gelled	Absorbed by the separator
Benefits	-Long service life (5 to 15 years) Lowest-cost technology	-Recombination => no water loss (no maintenance) -Very low gas release rate (safety)	
Disadvantages	-Water consumption (maintenance) Installation in specific areas (off-gassing)	-Shorter service life (specific failure modes) -More sensitive to temperature	

1.4. RECOVERY PROCESSES FOR LEAD-ACID BATTERIES

Used lead acid batteries can be recycled. [9], [8] or by regeneration [13], [16].

1.4.1. **Recycling**

The aim of recycling processes is to separate components into different fractions that can be reintroduced into a production process, treating hazardous fractions and minimizing process waste (effluents, emissions), energy consumption and costs. [3]. Many of the materials recovered from used lead-acid batteries (ULCBs) are used in the manufacture of new lead-acid batteries (LABs). For example, reclaimed lead is often used in the lead components of new BAPs, while reclaimed plastic can be used to make new cases. Or plastic that cannot be effectively separated from lead components in BAPUs is sometimes used as a complementary reductant in lead reduction and refining operations [9].

It is possible to recycle almost all the cells in a lead-acid battery. There are six (6) main stages in the recycling process [8] :

- Collecting and transporting batteries to a recycling plant ;
- Separating the various battery components ;

- ➢ Smelting and refining lead components ;
- ➢ Cleaning, then crushing or melting plastic components;
- ➢ Electrolyte purification and treatment (sulfuric acid) ;
- ➢ Waste treatment and disposal.

Figure 1.10 illustrates the typical process of a BAPU recycling plant.

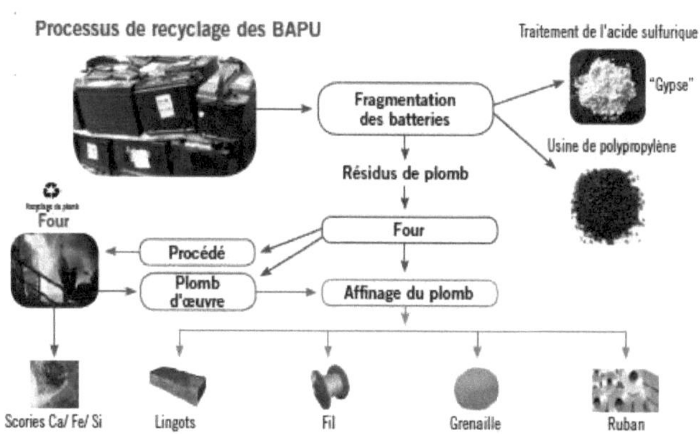

Figure *1.10*: Typical lead battery recycling process [9]

1.4.2. **Regeneration**

Regeneration here essentially concerns desulphation [13]. When a BAP is in the discharge phase, a lead sulfate deposit adheres to the lead plates. This whitish deposit increases the battery's internal resistance. The greater the battery's internal resistance, the greater the energy required to recharge the battery, until the battery becomes too sulfated and refuses to be charged. This sulfation is the cause of 80% of BAP failures. [16].

Regeneration (desulfation) can be carried out using three different processes [16]:

- ➢ The electrical process: Electronic regeneration exploits the battery's own resonant frequency: special equipment emits alternately positive and negative pulses at frequencies of between 1 and 10 kHz, to cause the lead sulphate crystals to "shatter". Once detached from the battery's lead plates, these crystals dissolve in the electrolyte, lowering the battery's internal resistance and making recharging easier.

- The chemical process: chemical regeneration involves adding a small quantity of magnesium sulfate (Epsom salt), or even better, sodium EDTA-4Na, to the battery electrolyte. These chemical compounds will dissolve the lead sulfate crust in the electrolyte. Treatment is generally performed by draining the electrolyte, then rinsing, then adding the "cleaning" solution, draining, then reinjecting the filtered electrolyte. It is also possible to add the chemical compound directly to the electrolyte without draining it (taking care not to exceed the electrolyte level).
- Combined process: this involves the combined use of both electrical and chemical processes on the battery to be regenerated. This process is more complex, since it involves the stages of both processes on the same battery.

1.5. TYPE OF FINANCIAL ANALYSIS

In finance, any activity requiring an initial investment is considered an investment project whose profitability must be assessed before any commitment is made, whether for potential income-generating or non-income-generating projects. [17].

For an investment project, the investment decision is based in part on a calculation that ensures that the anticipated future cash flows have an actuarial value greater than the cost of the resources required to carry out the investment. [18].

1.5.1. **Criteria for assessing the financial profitability of a project**

After estimating the various project cash flows (operating cash flow, changes in operating working capital requirements, fixed assets, terminal values, etc.), the discount rate and the project investment cost, the financial evaluation criteria are Net Present Value (NPV), Profitability Index (PI), Payback Period (PRP) and Internal Rate of Return (IRR). [18]. Below are the formulas for calculating the criteria used to assess project profitability [19].

- Cash flows

Cash flow is used to analyze a company's activity from a purely banking point of view, in order to highlight the company's net cash position. It is made up of cash inflows and outflows. Inflows are the sums that the company is expected to receive over the forecast period, in this case sales. Disbursements are the sums that the company will pay out over the forecast period, i.e. expenses and annual repayments to external investors. For each year n, we define Net Cash Flow (NCF):

$$FNT_n = EBITDA_n - Tax_n - ƛBFR_n - Investment_n + Subv_n \tag{1.15}$$

With

$EBITDA_n$ represents the project's gross operating surplus for year n ;

Income tax_n represents the corporate income tax charge;

$ƛBFR_n$ represents the change in working capital requirement in year n compared with year n-1 ;

$Investment_n$ represents the amount of investment financed in year n ;

$Subv_n$ represents the amount of subsidies received by the project company in year n.

- The NPV

NPV is the financial indicator that measures the profitability of an investment. NPV is used to assess the ability of an investment to create value over time, given the rate of return required by the company. Calculating NPV involves discounting the future cash flows generated by the project over its lifetime, at an appropriate discount rate:

$$VAN = \sum_{i=T0}^{N} FNT \text{ x } (1+K)^{-i} - CI \tag{1.16}$$

With :

CI : Investment cost ;

T_0 is the year in which the discounting calculation is performed;

N is the end year of the project;

k is the discount rate.

- THE IP

The profitability index compares the present value of cash flows generated by an investment project with the amount invested. A profitability index greater than 1 is equivalent to a positive net present value, and is an indicator of profitability, making it possible to measure the value created by each franc spent on the investment:

$$IP = (NPV / \text{initial investment cost}) + 1 \qquad 1.17$$

- DRCI

The payback period is the number of months or years required to recover an investment. The preferred investment is the one with the fastest payback period. The DRCI is generally determined by linear interpolation between two periods during which the cost of the investment will be equal to the forecast cash flow.

$$DRCI = \frac{CI - \text{FNT ACT cumul inf}}{\text{FNT ACT cumul sup} - \text{FNT ACT cumul inf}} + \text{AN FNT ACT cumul inf} \qquad 1.18$$

With :

CI: investment cost ;

FNT ACT cumul inf : Cumulative lower Net Cash Flow ;

FNT ACT cumul sup : Cumulative Net Cash Flow Discounted ;

AN FNT ACT cumul inf: Year of lower cumulative Discounted Net Cash Flow.

- SHOOTING

This is the minimum rate of return required for an investment project to be equivalent between its initial investment cost and its future cash flows. In other words, it's the rate for which the NPV is zero. It therefore measures the profitability of an investment.

This rate, which allows for equivalence between the cost of the investment and its future cash flows, can be calculated by simple mathematical resolution, by linear interpolation, or simply using the Excel spreadsheet (IRR function) of the following equation (assuming NPV = 0):

$$CI = \sum_{i=T0}^{N} \text{FNT x } (1+K)^{-i} \quad 1.19$$

With :

CI : Investment costs ;

$\sum_{i=T0}^{N}$ FNTi x $(1+K)^{-i}$ Discounted cash flows to be generated by the investment ;

K: IRR (to be researched) ;

N: estimated period over which the investment will generate cash flows.

It should be noted that this financial analysis will be carried out on the basis of a typical financial forecast model for an investment project [20].

CONCLUSION

Thanks to this study, we have understood the usefulness of energy storage systems, in particular electrochemical battery storage technologies. It should be noted that lead-acid technology is the oldest, the most suitable for many applications and the least expensive, making it the most widely used technology in the world in general and in Africa in particular.

CHAPTER 2: DEGRADATION OF LEAD-ACID BATTERIES AND POTENTIAL OF USED LEAD-ACID BATTERIES IN BENIN

INTRODUCTION

Like any piece of equipment in operation, a BAP can deteriorate with time or poor operating conditions.

This chapter is divided into three main sections:

- The first part presents the causes of BAP degradation;
- A second part will highlight the most frequent degradation modes of BAP ;
- The final part of this chapter assesses the potential of BAPUs in Benin.

2.1. CAUSES OF DETERIORATION

The main causes of lead-acid battery degradation are numerous. May et al cited fifteen (15) as follows [11], [13], [21] :

2.1.1. Corrosion of the positive grid

The positive grid is maintained at charging voltage, immersed in sulfuric acid, and will corrode throughout the battery's life when maximum charging voltage is reached. For high corrosion resistance, the grid alloy is either lead-antimony, lead-calcium-tin, lead-tin or pure lead, the grid thickness and other grid design parameters are selected to provide sufficient metal for the expected life of the battery. [21].

The grid manufacturing process and the microstructure of the grid alloy are important. Grid corrosion is accelerated by higher charging voltages. Voltage is set to obtain a fully charged battery without excessive water loss, and corrosion is kept to a level that will achieve the expected service life. Correct setting of charging voltage is essential [22]. Grid corrosion is also temperature-sensitive. Grid resistance increases over the life of the battery, accelerating towards the end of life. There will also be loss of connectivity between the grid and the active material [21].

2.1.2. Positive network growth

Variations in the volume of active material and the volume of corrosion product exert stress on the grids, which can deform as the battery ages. Grids can grow and come into contact with the negative group bar, causing short-circuits. They can also grow so that they are no longer insulated at the edges of the plates [21].

Some battery models safely accommodate this phenomenon by using an insulator under the group bar. Others allow the grids to grow downwards or to accommodate a certain distortion. This can lead to pressure on the pillar seal, which can become distorted and develop leaks. The housing may crack and start to leak. Grid growth also leads to a loss of connectivity between the grid and the active material, increasing internal resistance and reducing battery capacity. Grid growth can be reduced by correctly selecting the grid alloy and ensuring that grid thickness and wire cross-sections are adequate [21].

Tubular positive plates are generally more resistant to the effects of grid growth than bonded plates, as lateral growth is limited by fabric gauntlets and vertical growth can be reduced. Lead-antimony alloys are more resistant to grid growth than lead-calcium-tin alloys because they have higher tensile and creep strengths, but for VRLA batteries, lead-calcium-tin, lead-tin or pure lead should be used for grids to suppress water losses [21].

2.1.3. Sulfation

The normal discharge product on positive and negative plates is lead sulfate. This is normally very finely divided and easy to recover by recharging, but with time and cycling, it tends to agglomerate and becomes more difficult to recharge, eventually leading to a loss of capacity. There will be an increase in the cell's internal resistance and a loss of performance. Sulfation will be increased if the battery is left in a state of partial or total discharge for long periods of time [21].

Sulphated batteries have a high internal resistance and can only deliver a small fraction of the normal discharge current. Sulfate also affects the charge cycle, resulting in longer charge times, less efficient and incomplete charging, and higher battery temperatures [11], [13].

2.1.4. Softening of the active ingredient

Over time, the active material may become soft and less cohesive. This phenomenon will be more rapid if the battery is subjected to deep cycling, and will result in a loss of capacity. For positive plates, this phenomenon can be reduced by using higher-density pastes [21].

2.1.5. Acid stratification

During recharging, sulfuric acid is produced from both plates as lead sulfate is reduced on the negative plate and oxidized on the positive plate, and the acid with its higher concentration and hence density tends to move towards the bottom of the cell. The acid is stratified with a density gradient from the top to the bottom of the cell. This increases the tendency for sulfation to occur in areas of higher acid concentration, and also leads to variations in active ingredient utilization. It can be eliminated by simple acid recirculation pumps. This problem affects open cells, and VRLA cells are less affected by stratification, as the AGM separator or gel immobilizes the electrolyte. For AGM types, the wick height of the separator is limited, which limits the height of the cell. High cells are generally operated from the side to avoid any problems of acid density variation [21].

2.1.6. Drying

For VRLA batteries at end-of-life or under certain fault conditions, such as excessive charging voltages, the cells lose water, resulting in separator shrinkage, loss of compression and reduced contact between plates and separator. These conditions lead to a loss of capacity and an increase in internal resistance. Higher operating temperatures also increase the risk of drying out. If desiccation occurs more rapidly, for example as a result of high applied voltage, it can lead to thermal runaway. It can also occur over extended periods at normal charging voltages, but the battery would have exceeded its normal service life. Cells are designed in such a way that desiccation is not a normal failure mode. Desiccation may be the result of a mechanical fault, high ambient temperature, incorrect charging or a combination of these contributing factors, or it may occur in a battery that has exceeded its end-of-life. If charging voltage is correctly controlled, drying out should not be a failure mode. For open batteries, there is no risk of drying out with regular water additions for maintenance. [21].

2.1.7. Leaking pillar joints

Pillar seals may leak in service due to manufacturing defects. VRLA cells can dry out, and acid on the battery surface can cause ground fault currents, which in turn can lead to overheating and thermal runaway. [21].

2.1.8. Cover gasket leak

Cover gaskets can also leak due to manufacturing defects, with results similar to those of the pillar gasket. Good manufacturing practices are the key to avoiding this cause of failure [21].

2.1.9. Vent failure

VRLA battery vents can be missing, ejected or fail to close, resulting in reduced capacity, accelerated drying, electrolyte loss or both. For open cells, no damage occurs if the vents are damaged [21].

2.1.10. Mechanical damage

Accidental mechanical damage can cause cells to leak, leading to a failure similar to that of a pillar seal leading to a failure similar to that of a pillar seal leak. Loss of electrolyte can lead to drying out and loss of capacity [13], [21].

2.1.11. Group bar corrosion

The connection between the group bar and the plate lugs may be corroded and eventually disconnected. The splice bar alloy and the connection between the group bar and the plate lugs must be correctly specified, especially if this is a manual operation. [21].

2.1.12. Internal short circuit

They may result from damage or penetration of the separator. These can be hard shorts leading to rapid failure, or soft shorts where the fault is relatively minor and may not be apparent at first, as it can develop over time. Internal short circuits lead to loss of capacity through self-discharge before use. Shorts can lead to thermal runaway in VRLA cells [21].

2.1.13. External hydrogen ignition

Batteries emit small quantities of hydrogen during charging. Under normal operating conditions and in a well-ventilated enclosure, this quantity should not reach the flammability limit of hydrogen in air, but it may do so in the event of a fault. [21]. Fireproof vents must be used.

2.1.14. Overheating of external connections

If inter-cell or inter-monobloc connections are not secured, there is a risk of overheating, particularly during discharge, due to the high local resistance and, in extreme cases, this can lead to fire. [21].

2.1.15. Thermal runaway

Thermal runaway in VRLA batteries is an unstable condition where the application of charging voltage causes the battery temperature to rise uncontrollably, and in extreme cases this can lead to a battery fire or explosion. Current is not normally limited, as the battery will be able to absorb large currents for recharging after a discharge cycle[21].

There are a number of causes of thermal runaway, as indicated above: drying out, internal short-circuiting, cell inversion and other faults. It can also be caused by excessive load voltage accelerated by rising temperatures in the local environment. For open cells, there are no difficulties [21].

NB: According to May et al (2018) [21]some of the causes of failure described can be avoided by best practice in battery design, manufacture and operation [22] but others, such as positive grid corrosion and growth, sulfation and softening of active materials require a combination of better materials, further improvements in battery design and careful control of charging parameters.

2.2.MOST FREQUENT MODES OF DAMAGE

ADEME, in its report on the state of the art of desulphation technologies (2011, which brought together more than 13 BAP regeneration professionals) defined the main and most frequent degradation modes of BAPs, all technologies combined, according to the causes cited above. Table 2.1 shows these degradation modes.

Table 2.1: Most frequent degradation modes on lead-acid batteries *[13]*

Type of damage	Causes	Consequences	Battery type concerned	Reversibility
Corrosion of positive electrode	- Spontaneous oxidation at rest and under load - Accelerated by electrolyte stratification	- Short circuits - Impaired electrical conduction	- Main cause for stationary batteries - Secondary cause for traction batteries	No
Active ingredient degradation	Spontaneous phenomenon	Short circuits	Main cause for traction batteries	No
Electrolyte drying	- Gas recombination never 100% efficient - Water loss during overloads	Accelerated sulfation	Information not available [13] VRLA technology [21]	No
Sulfation	- Incorrect use or maintenance - Accelerated by electrolyte stratification	Impaired electrical conduction	Secondary cause for traction and stationary batteries (due to misuse: discharge time too long, undercharging, etc.).	Yes : Desulfation

The various degradation modes mentioned in the table are defined as follows [13] :

2.2.1. Corrosion of the positive electrode

It corresponds to the spontaneous oxidation of the lead on the positive electrode grid when the battery is at rest and charging.

2.2.2. Degradation of the active ingredient

This is the decomposition of the active material in the positive electrode into particles that accumulate at the bottom of the battery. The particles can penetrate the porous networks of the separators and accumulate between two plates.

2.2.3. Sulfation

This is a build-up of lead sulfate in the battery. During discharge, lead sulfate crystals ($PbSO_4$) form on the positive and negative electrodes. This phenomenon normally disappears during recharging. However, under certain conditions (prolonged or excessively deep discharge, high temperature, gasification of the electrolyte, undercharging), stable islands of lead sulfate appear and are no longer dissolved during recharging.

NB: After all these degradations, there is another parameter that seriously affects battery life: its operating temperature, which can be a function of ambient temperature. Table 2.2 below gives an example of the effect of operating temperature on battery life.

Table 2.2: Effect of temperature on battery life *[1]*

Average temperature	AGM Deep cycle Years	Gel Deep cycle Years	Gel Long service life Years
20°C / 68°F	7 - 10	12	20
30°C / 86°F	4	6	10
40°C / 104°F	2	3	5

The rule of thumb is that shelf life decreases by 50% for every 10°C above 20°C. *At* 30°C, service life is reduced by 50%, and at 40°C by a further 50%.

As a result, you will have a battery life of a quarter of what it would be at 20°C, hence the poor experience of battery life in sub-Saharan Africa, because a battery can become hot (increase in temperature) for the following reasons:

- Rapid discharge;
- Fast charging ;

- A warm environment.

2.3.ASSESSING THE POTENTIAL OF BAPU IN BENIN

This assessment is based on our various achievements in the field of renewable energies and photovoltaic solar power in particular.

In order to identify the prospects for localizing the development of solar energy, we need to present the main government initiatives in this field, most of which have been carried out through major infrastructure projects. Drawing on the experience of its first public initiatives (a project to build 24 solar villages in 1990), the Beninese government has, in recent years, re-launched two new solar energy development projects [23]. These are :

Firstly, the Regional Program for the Development of Renewable Energy and Energy Efficiency (PRODERE), which began in June 2014, is a program co-financed by the West African Economic and Monetary Union (UEMOA) and the Beninese state, which aims to substantially reduce the energy deficit in UEMOA member countries. It focuses on a few priority actions (UEMOA and SABER, 2013):

- The installation of solar street lamps on university campuses and main thoroughfares in four major cities;
- The construction of solar micropower plants in rural areas;
- The installation of solar photovoltaic kits in selected Beninese homes, schools and health centers in rural areas.

Secondly, following PRODERE's initial conclusive results, the Beninese government set up its own renewable energy development project called PROVES.

PROVES (Projet de valorisation de l'énergie solaire), led by ANADER (Agence Nationale pour le Développement des Energies Renouvelables). This second major project aimed to :

- Solar street lighting with the installation of 15,000 lampposts nationwide;
- Construction of solar photovoltaic micropower plants in 105 rural localities.

Our assessment therefore focuses on the numbers of solar street lighting and solar photovoltaic micropower plants already in place, among those planned for these two major projects.

2.3.1. Current status of solar street lighting in Benin

Table 2.3: Status of solar street lighting in Benin in 2020 [24]

N°	Department	Ministry of Energy projects			Total Ministry of Energy projects	Other projects (town halls and road projects)	Total streetlights installed	Floor lamps being installed
		PROVES	PRODERE	PILAKS-PV				
1	ALIBORI	880	-	50	930	30	960	0
2	ATACORA	1 120	40	10	1 170	56	1 226	0
3	ATLANTIQUE	2 345	82	750	3 177	1 413	4 590	669
4	BORGOU	1 870	95	30	1 995	645	2 640	0
5	COLLINES	828	10	400	1 238	24	1 262	0
6	COUFFO	870	0	0	870	64	934	0
7	DONGA	570	0	40	610	236	846	0
8	LITTORAL	2 666	420	0	3 086	0	3 086	321
9	MONO	915	0	200	1 115	125	1 240	800
10	OUEME	1 759	63	20	1 842	70	1 912	811
11	PLATEAU	540	10	0	550	0	550	0
12	ZOU	1 347	10	0	1 357	180	1 537	0
Total		15 710	730	1 500	**17 940**	**2 843**	**20 783**	**2 601**

Proportion of streetlights installed per project	76%	3%	7%	**86%**	**14%**	**100%**

Table 2.3 shows the number of streetlights installed per project in each of Benin's departments from 2015 to 2020. It also shows the number of streetlights being installed by the Ministry of Infrastructure and Transport (MIT) from 2020.

A total of **20,783 solar streetlights** have already been installed in Benin, and **2,601** are currently being installed by the Ministry of Infrastructure and Transport as part of roadworks.

Of the 20,783 solar streetlights already installed, 17,940 (or 86% of the total installed) come from Ministry of Energy projects (PROVES, PRODERE and PILAKS-PV) and the remainder (2,843 or 14%) from communal and road development projects. [24]. The breakdown by project is as follows

- 15,710 streetlights installed for the PROVES project (76%) ;
- 1,500 for PILAKS-PV (7%) ;
- 730 on behalf of PRODERE (3%) ;
- 2,843 streetlights installed for municipal and road improvement projects (14%).

2.3.1.1. Evaluation of materials required for the rehabilitation of solar lighting infrastructures

The last streetlight installations for both projects were handed over in 2015. Given the average life of batteries, which is only around 4 to 5 years [25]the replacement of all the batteries in these installations has already been scheduled for 2020.

In view of the above, the equipment required for the rehabilitation of solar street lighting infrastructures has been evaluated and summarized in Table 2.4. [24] below:

Table 2.4: Evaluation of equipment needed to rehabilitate solar street lighting infrastructures

Designation	Quantity	Rehabilitation materials
Functional floor lamps	12 328	Batteries to renew
Faulty street lamps	6 804	Batteries and regulators except those from PROVES
Faulty street lamps in the PROVES project	3 947	
Faulty street lamps, except for those in the PROVES project	2 857	Batteries and Regulators
Streetlights vandalized, except for those on main roads	1 584	All in One
Vandalized street lamps on main roads	67	Batteries and Regulators

In terms of batteries to be regenerated, there is a potential of 19,132 batteries (6,804 batteries from defective streetlights and 12,328 batteries from functional streetlights that are no longer providing their services, and are therefore at the end of their first life).

These streetlights are equipped with 12V / 150Ah gel batteries, and our first potential number of batteries to be regenerated is **19,132 12V / 150Ah gel batteries** by 2020.

2.3.2. Current status of PV micropower plants in Benin

According to the report by ESEIM (Energy System Equipments Installation & Maintenance, in May 2019), a firm hired by ECREEE (ECOWAS Centre for Renewable Energy and Energy Efficiency) to carry out inventories of PV micropower plants installed in Benin, as regards the state of the art, we can conclude that [26] :

- By 2019, a total of 80 solar PV micropower plants had already been installed in 78 localities in Benin, including 74 by PROVES and 6 by PRODERE.

Tables 2.5 and 2.6 show the distribution of these micropower plants by department and size.

Table 2.5: Number of micropower plants by départment

N°	Department	PRODERE	PROVES	Total
1	Atacora	3	12	**15**
2	Alibori		11	**11**
3	Borgou	1	9	**10**
4	Donga			**0**
5	Hills	1	6	**7**
6	Zou	1	16	**17**
7	Couffo		8	**8**
8	Tray		2	**2**
9	Mono		5	**5**
10	Atlantic		3	**3**
11	Ouémé		2	**2**
	BENIN	**6**	**74**	**80**

Table 2.6: Number of mini-power stations by size

N°	Size	PRODERE	PROVES	Total
1	15 kWp	2	0	**2**
2	20 kWp	1	13	**14**
3	30 kWp	1	26	**27**
4	40 kWp		35	**35**
5	45 kWp	2	0	**2**
	BENIN	6	74	**80**

2.3.2.1. Inventory of battery fleets

Generally speaking, the batteries installed in both projects are **OPzV 2 V / 2000 Ah.** [26]. The number of batteries installed as a function of micropower plant size is given in Table 2.7 [27].

Table 2.7: Number of batteries installed by micropower plant size

N°	Size	Number of 2V / 2000 Ah batteries per size	PRODERE	PROVES
1	15 kWp	24	2	0
2	20 kWp	44	1	13
3	30 kWp	74	1	26
4	40 kWp	96	0	35
5	45 kWp	120	2	0
Total batteries installed			**406**	**5856**

Table 2.7 above shows that the batteries installed in the two PRODERE and PROVES projects are **406** and **5856** respectively, for a total of **6262 OPzV 2 V / 2000 Ah** batteries.

The PRODERE and PROVES plants were handed over in 2017 and 2018 respectively, and are currently in operation for 5 and 4 years respectively.

Given that battery life is no more than 4 to 5 years, the batteries in these plants will already be at the end of their first life in 2022.

In terms of the potential currently available (2022) at micropower station level, we only have **6262 batteries**. This potential may seem a little insignificant, but it's already enabling us to set up a business model.

As a result of this assessment, the quantified BAPU potential in Benin is shown in table 2.8 below.

Table 2.8: BAPU potential in Benin

BAPU potential in 2O22

Batteries	Quantity
12 V / 150 Ah (Floor lamps)	19 132
2 V / 2000 Ah (Micro power plants)	6262

However, when we take into account the non-homogeneity of regeneration criteria on batteries (mechanical damage, satisfactory voltage, electrolyte level and density, age, capacity and short-circuit fault), this potential cannot be considered fully exploitable, which means that we're looking at around 80% of regenerable batteries for a given quantity. [13].

This 80% rate applied to our first potential gives us the potential of regenerative batteries with which to build our business model. This potential is shown in table 2.9 below.

Table 2.9: Potential of regenerable BAPU

Potential for regenerable BAPU	
Batteries	**Quantity**
12 V / 150 Ah (Floor lamps)	15 305
2 V / 2000 Ah (Micro power plants)	5009

NB: For the financial analysis of the project's profitability, the potential taken into account is that of regenerative batteries, and the analysis is based on a total potential of 20,000 BAPUs, with 15,000 BAPU gels 12 V / 150 Ah and 5,000 BAPU OPzV 2 V / 2000 Ah.

CONCLUSION

This chapter describes the causes and modes of failure that can occur during the service life of BAPs. They can be classified according to mechanical, electrical and physicochemical causes of failure.

Among physicochemical degradation modes, corrosion and degradation of the active material are respectively the main causes of stationary battery failure, since these phenomena are irreversible. Of all the types of physicochemical degradation that can affect BAPs, only sulfation is reversible.

An assessment of the potential for BAPUs in Benin through government action has enabled us to quantify a total of **25,934** BAPUs currently available (2022), with significant potential in the very near future. This potential is divided into two types as follows: **19,132 Gel 12 V / 150 Ah BAPUs** and **6,262 OPzV 2 V / 2,000 Ah BAPUs.**

What's more, with Benin more than ever on the right track in its energy transition, developing its renewable energy potential and optimizing energy efficiency, the potential of BAPU can grow exponentially in the very near future, especially with the implementation of new projects planned by the Ministry of Energy under the 2021 - 2026 Government Action Plan.

CHAPTER 3: TECHNICAL STUDY OF LEAD BATTERY REGENERATION

INTRODUCTION

The aim of this study is to take stock of existing technologies for BAPU desulphation (also referred to as "regeneration" by industry professionals), their operating modes, and a comparative analysis of the different processes from a technical, economic and environmental point of view, in order to select the best process.

BAP desulfation technologies are designed to extend the useful life of BAPU by combating sulfation.

This chapter is structured in three parts as follows:

- The first part presents a general overview of regeneration;
- The second part reviews the state of the art in BAPU regeneration processes;
- The third section compares the different processes.

3.1. GENERAL INFORMATION ON REGENERATION

3.1.1. **Why desulphation?**

- Sulfation is a natural and reversible phenomenon, accelerated in particular by heat [28] and deep discharge (in the case of solar batteries);
- This sulfation is the cause of 80% of BAP failures due to loss of capacity. [16] ;
- A heavily sulfated battery will be unable to perform its service and will need to be replaced or revitalized;
- A battery is irreparable when it becomes short-circuited (excess lead sulfate leading to perforated insulation).

3.1.2. **The benefits of regeneration**

The usefulness of BAPU regeneration is summarized in figures 3.1 and 3.2 below.

The main stages in the desulphation-free life cycle of a battery can be summarized as follows:

Figure 3.1: Life cycle of a non-regenerated battery

When a desulphation step is included, the life cycle of a battery is modified as follows:

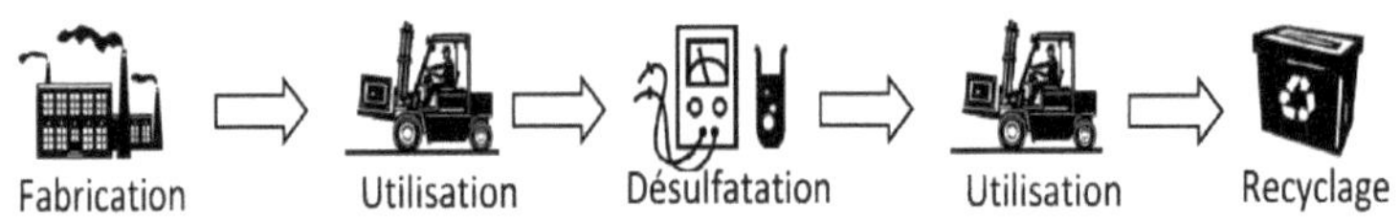

Figure 3.2: Regenerated battery life cycle

Not to mention the time savings involved in procuring new batteries for replacement, regeneration is therefore an ideal solution, providing an immediate economic and ecological solution for all concerned.

3.1.3. **Definition**

Those involved in desulphation define it as a process for extending the useful life of BAPUs or batteries whose performance no longer meets the user's needs, by remedying the phenomenon of sulphation. The same process can sometimes be applied to new (or near-new) batteries, to act preventively against the sulfation phenomenon [13].

This BAPU desulphation activity is a market that is currently being developed for both industrial batteries (traction or stationary batteries) and automotive batteries (starter batteries).

3.1.4. **Regeneration principle**

Table 3.1 below provides an overview of the regeneration operating principle.

Table 3.1: Regeneration overview

Regeneration principle	Method	Delivery of controlled, high-power electrical, chemical and/or combined pulses
	Duration	From 20 hours for starter batteries to 5 days for industrial batteries
	Effect	Pulses break up lead sulfate crystal lattice and reconstitute lead plates
	Results	Battery restored to original capacity: New Life
Possible for which battery types?	Starting, pulling, stationary	GEL, AGM, with or without maintenance, etc...
Relevant for which applications?	Transport	Cars, trucks, vans, boats, etc.
	Construction equipment	Excavators, cranes, cherry pickers, scrubber-driers...
	Handling	Pallet trucks, forklifts, stackers, etc...
	Energy storage	UPS, solar panels, wind turbines, etc...

3.2. STATE-OF-THE-ART BATTERY REGENERATION PROCESSES

Three desulphation (regeneration) processes were identified in the course of this study. They are as follows:

- Chemical process ;
- Electrical process ;
- Combined electrical and chemical process.

The operating modes of these three processes are described below.

3.2.1. **Chemical regeneration process**

This process involves a chemical component only. The chemical process consists of adding a small amount of organic polymer to the battery electrolyte. [29]magnesium sulfate, EDTA, or hydrogen peroxide [30]among others) in liquid, powder or capsule form. These chemical compounds will dissolve the lead sulfate crust in the electrolyte.

The disadvantage of this process lies in the complex determination of the specific volume of additive to be added, which varies according to the characteristics of the battery to be regenerated, and in the handling of toxic and hazardous chemicals.

3.2.1.1. Chemical process efficiency

The effectiveness of this process has been verified by several authors who have desulfated BAP using chemical additives. These include :

- Minami et al (2004) [31]Ikeda et al (2007) [29] and Ikeda et al (2008) [30]After eight years (8 years) of research (1996 to 2004) at the International Technology Exchange's Battery Research Institute (ITE Battery Research Institute), they had succeeded in finding chemical additives to effectively desulfate BAPU ;
- ADEME [13] in its state-of-the-art report on desulphation technologies three (3) professionals use this process (2011);
- Paglietti [32] in the study evaluating the effect of additives in the electrolyte to determine the maximum permissible concentration for regenerating BAPs.

3.2.1.2. Chemical process diagram

Treatment is generally performed by draining the electrolyte, then rinsing, then adding the cleaning solution, draining, then reinjecting the filtered electrolyte. It is also possible to add the chemical compound directly to the electrolyte without draining it (taking care not to exceed the electrolyte level).

As for the professionals, each has his or her own way of doing things, but all have points in common, from which we have drawn the following synoptic diagram.

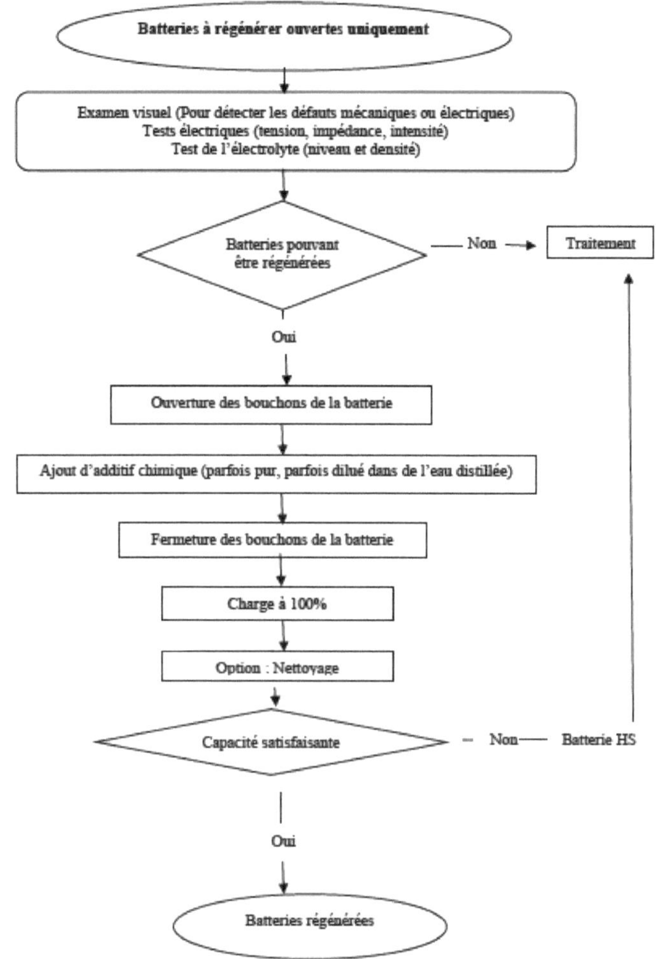

Figure *3.3*: Typical chemical process flow diagram for battery regeneration [13]

During preparation, a visual inspection of the battery is carried out, to ensure that it can be regenerated (desulphated). Batteries showing signs of leakage, corrosion, damaged terminals or other visible, irreversible damage should be eliminated from the outset. Voltage, impedance, current and electrolyte density tests complete the preparatory phase.

Desulphation (regeneration) as such can then begin: the players inject a chemical additive into the electrolyte. The quantity injected is calculated according to the battery's parameters (mainly its capacity). Distilled water can also be added to adjust the electrolyte level.

To complete desulphation, the battery is charged to 100%. This activates the additive and allows the battery to be reused.

3.2.1.3. Description of a chemical regeneration plant

The complete regeneration process follows the three stages mentioned above (preparation, desulphation and finalization), during which various pieces of equipment are used.

Table 3.2 summarizes the installation of the chemical regeneration process, specifying the various stages, the equipment used and the technical characteristics of the process.

Table *3.2*: Characteristics of a chemical regeneration plant

Eligibility criteria for regeneration	
No mechanical damage	
Satisfactory voltage	
Electrolyte level and density satisfactory	
Age (not too old) and capacity (not too low)	
Battery types concerned	
BAP (open only)	Starting, pulling and stationary
Different stages and equipment used	
Preparation	Multimeter, voltmeter, ammeter, impedance meter
Desulfation	Organic polymer, EDTA, hydrogen peroxide
Finalization	Classic charger or special charger, cleaning device
Technical data	
Process time	A few hours, results after several charge/discharge cycles
Use of the process in preventive maintenance	Possible
Operator	An operator is needed at every stage

Control criteria for verifying success	Regenerated battery capacity
Success rates	70% to 80% of BAP accepted for regeneration
Service life of regenerated BAP	Service life can be multiplied by 1.5 or 2
Regenerative capacity	85 to 95% of manufacturer's capacity

3.2.2. **Electric regeneration process**

The process involves an electrical component only. Electrical regeneration exploits the battery's own resonant frequency. A special device called a regenerator (desulfator) emits pulses of varying frequencies, alternately positive and negative, to break up the lead sulfate crystals. Once detached from the battery's lead plates, these crystals will dissolve in the electrolyte, lowering the battery's internal resistance and making recharging easier.

The disadvantage of this method is that the pulse frequency cannot be fixed, as the battery's natural resonance frequency will change as the regeneration process progresses. Electronic processing equipment must therefore be able to identify the natural resonance frequency of the battery being treated at any time, and adjust pulse frequencies accordingly - hence the interest in the latest BAP regenerators.

3.2.2.1. Electrical process efficiency

The effectiveness of this process has been verified, approved and validated after several research, improvement, correction and accompanying patent applications by professionals and manufacturers of electrical regeneration devices. All these studies have proved the possibility of significantly extending the service life of the batteries to which these devices are connected. They include the work of :

- Lam et al [33]who demonstrated the advantages of a pulsed current load of the order of 8 A over the conventional invariable current load of the order of 0.75 A (1995);

- Minami et al [34]from his experiments on 28 Ah vehicle starter batteries, concludes that pulse charging extended the possible discharge time of his batteries (2004);
- Yan Zhang et al [35]who demonstrated the performance of a high pulse current regenerator on which a BAP is discharged and then recharged at high pulse current of the order of 5 to 10 C or 250 to 500 A (2008);
- Mathew et al [36]who report desulfating BAPs using a desulfation current regenerator comprising a repeating pattern including an ON pulse of about 0.75 ms followed by an OFF period of about 4.5 ms, which can be applied to the battery at an operator-adjustable peak amperage of about 0-350 amperes (2010) ;
- ADEME [13]in its report on desulphation seventeen (17) professionals in the field confirm that they use the electric process because of its efficiency (2011);
- Oumar [37]method and device for diagnosing the regenerability of a battery (2015);
- Jamratnaw [11]BAP desulfation using high-frequency pulses (2017);
- Ibrahim et al [38]who restored the capacity of a flooded industrial forklift BAP (840 Ah, 36 V) from 68% to 99% using a high pulse frequency desulfator producing a maximum direct current (DC) of 500 A and instantaneous heat of 27° C to 48° C to dissolve the $PbSO_4$ on the plates (2020) ;
- Ohajianya et al [39]To support US patent applications for certain players, he desulfated four 100 Ah gel-valve BAPs with a high-frequency pulse regenerator. In conclusion, his experience has enabled him to understand that the best regenerators are those that are chargers and incorporate the mechanism to prevent overcharging (which are therefore able to identify the battery's natural resonance frequency at all times) than those that are not and operate with separate chargers (2021) ;
- Marly Diallo [40]CEO of BRT Energy (expert in BAPU regeneration in Nigeria, Ghana, Rwanda and Ethiopia) in her interview with BBC Afrique confirms the use of the electrical process to regenerate BAPU (2021);

- Battery more [41](supplier of BAPU regeneration machines and services) in its database counts over three hundred and fifty (350) electrical regeneration machines sold and installed worldwide, including some in Africa (Ivory Coast, Senegal, Burkina Faso, Chad etc.....).

3.2.2.2. Electrical process diagram

Treatment is essentially by connection to a regenerator. According to the professionals, the following synoptic can be retained.

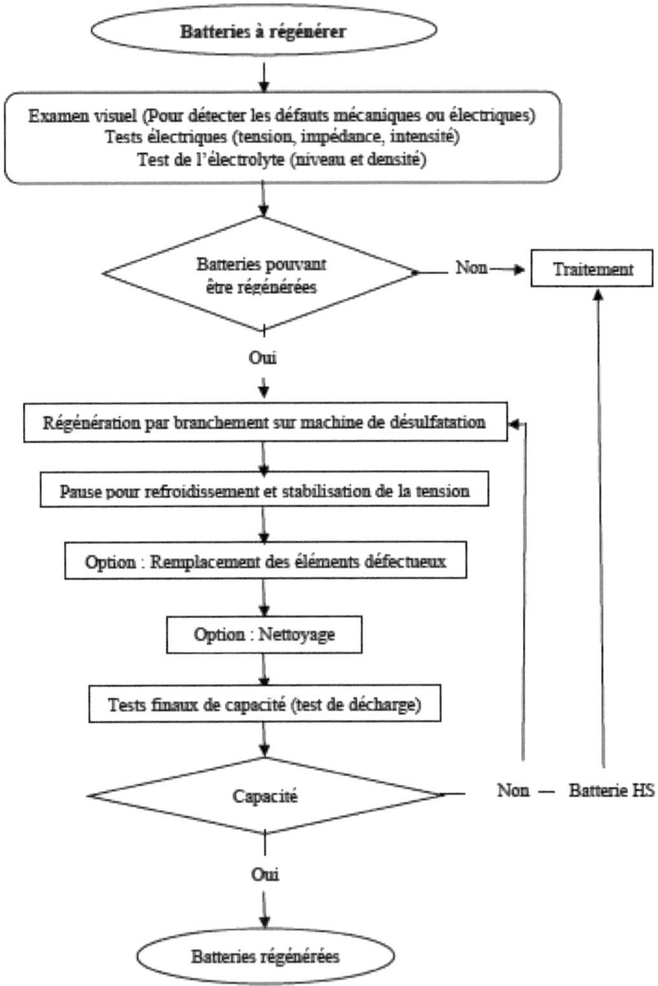

Figure *3.4*: Typical electrical process diagram for battery regeneration [13]

The preparatory stages are identical to those for the chemical process, and once this preparation has been completed, the battery can be desulphated.

Regeneration is triggered by the application of electrical pulses. The intensity of these pulses varies from player to player: most often strong (120-300 A), others weak (8 A).

After desulphation, the battery is cooled and its voltage stabilized. Any damaged and identified components are replaced, if not already done so. Final checks are then carried out, including a discharge test (to check the capacity of the regenerated battery for industrial batteries), as well as voltage, current and electrolyte level tests.

3.2.2.3. Description of an electrical regeneration plant

The electrical regeneration plant is described in table 3.2 below. This is a complete description from receipt of the BAPs through to final regeneration, taking into account the various stages and the equipment used.

Table *3.3*: Characteristics of an electrical regeneration plant

Acceptability criteria for regeneration	
No mechanical damage	
Satisfactory voltage	
Electrolyte level and density satisfactory	
Age (not too old) and capacity (not too low)	
Battery types concerned	
All types of BAP (open and sealed)	Starting, pulling and stationary
Different stages and equipment used	
Preparation	Multimeter, voltmeter, ammeter, impedance meter, densitometer, discharge tester, specific starting current tester, thermometer
Desulfation	Regenerator, computer, connecting cable
Finalization	Discharge bench, connectors, multimeter, impedance meter, densitometer, cleaning device (optional)
Technical data	
Working environment	Sheltered, ventilated area maintained between 18 and 21° C
Process time	3 hours to 96 hours
Use of the process in preventive maintenance	Possible
Operator	An operator is needed at every stage

Control criteria for verifying success	Regenerated battery capacity
Success rates	80% to 100% of BAP accepted for regeneration
Service life of regenerated BAP	Usage time can be multiplied by 2
Regenerative capacity	95 to 100% of manufacturer's capacity

3.2.3. Combined regeneration process

This process uses a combination of electricity and chemicals. A battery containing chemical additives is connected to a regenerator which emits electrical pulses of varying frequencies.

When the regenerator is able to determine the battery's natural frequency, the disadvantage of this process is still determining the specific volume of chemical additive to be injected.

3.2.3.1. Combined process efficiency

The effectiveness of this process has been proven by an experimental study dating from 2005 and two professional publications from 2011 and 2012. They are respectively:

- Ikeda et al [42] in his work, reports that on one of his low pulse current desulfators (about 30 A for a 30 AH battery) the efficiency was higher when connected to a battery filled with chemical additive supplied by ITE (2005) ;
- ADEME [13] in its report on desulfation nine (9) professionals in the field confirm that they use the combined process (2011);
- GREENKRAFT expertise [16] (BAP regeneration professionals) in its practical publication confirms that the combined process is also of proven effectiveness (2012).

3.2.3.2. Combined process diagram

Depending on the player involved, treatment takes place in two or even three stages: injection of the chemical additive, connection to the regenerator and connection of an electronic box to the battery terminals. This process is described in figure 3.3.

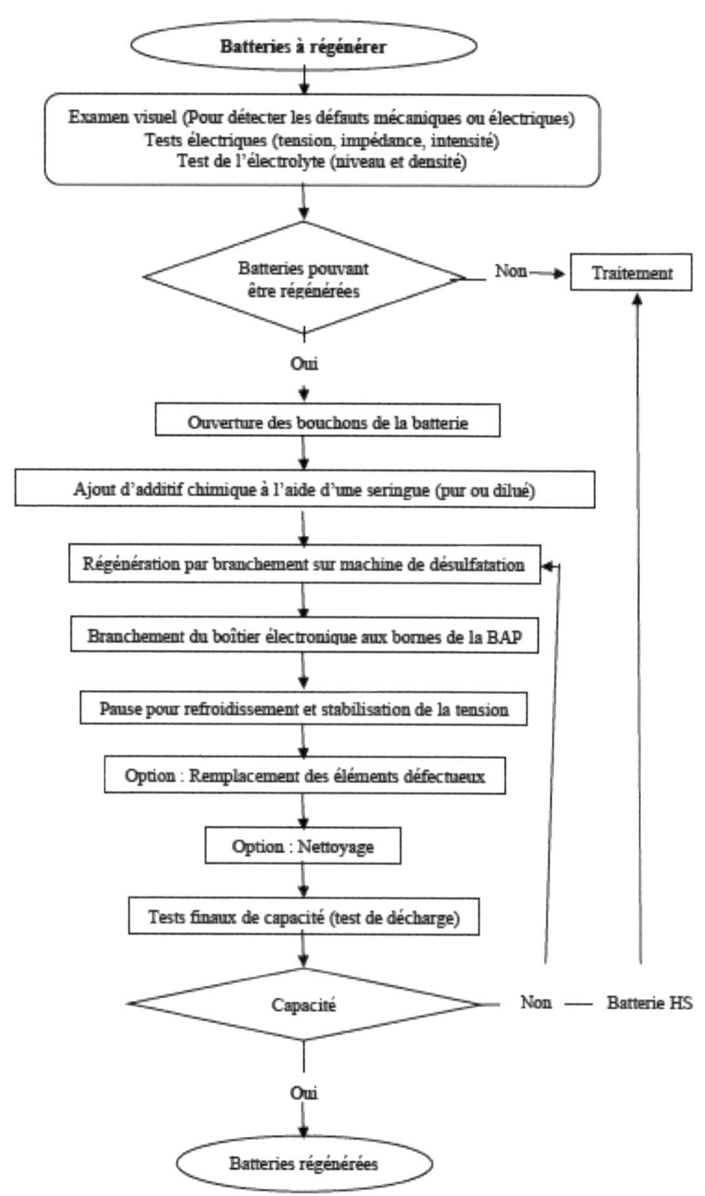
Batteries à régénérer
Examen visuel (Pour détecter les défauts mécaniques ou électriques)
Tests électriques (tension, impédance, intensité)
Test de l'électrolyte (niveau et densité)
Batteries pouvant être régénérées
Non
Traitement
Oui
Ouverture des bouchons de la batterie
Ajout d'additif chimique à l'aide d'une seringue (pur ou dilué)
Régénération par branchement sur machine de désulfatation
Branchement du boîtier électronique aux bornes de la BAP
Pause pour refroidissement et stabilisation de la tension
Option : Remplacement des éléments défectueux
Option : Nettoyage
Tests finaux de capacité (test de décharge)
Capacité
Non
Batterie HS
Oui
Batteries régénérées

Figure 3.5: Typical block diagram of a combined battery regeneration process [13]

The various preparation stages are identical to those for the chemical and electrical processes: visual inspection, cleaning, followed by electrical testing and electrolyte density testing (depending on battery type).

The special feature of this combined process is that it offers desulphation in two or even three stages (depending on the manufacturer). First, a chemical additive is injected into the battery. Distilled water and/or electrolyte is sometimes added after opening the battery caps. Secondly, the battery is connected to a desulphation machine. Thirdly, some manufacturers connect an electronic box to the battery terminals, which continuously sends out electrical impulses. For professional users, this box is an integral part of the desulphation process.

After a cooling period, tests identical to the initial tests are carried out at the end of the process. A discharge test is routinely carried out for industrial batteries.

3.2.3.3. Description of a regeneration plant using the combined process

A typical installation for this process is described in table 3.4 below.

Table *3.4*: Characteristics of a combined regeneration plant

BAP eligibility criteria for regeneration	
No mechanical damage	
Satisfactory voltage	
Electrolyte level and density satisfactory	
Age (not too old) and capacity (not too low)	
Battery types concerned	
All types of BAP (mostly open)	Starting, pulling and stationary
Different stages and equipment used	
Preparation	Multimeter, voltmeter, ammeter, impedance meter, densitometer, discharge tester, specific starting current tester, thermometer
Desulfation	Chemical additive, regenerator, computer, connecting cable, electronic box (optional)

Finalization	Discharge bench, connectors, multimeter, impedance meter, densitometer, cleaning device (optional)
Technical data	
Working environment	Sheltered, ventilated area maintained between 18 and 21° C
Process time	4 hours to 48 hours
Use of the process in preventive maintenance	Possible
Operator	An operator is needed at every stage
Control criteria for verifying success	Regenerated battery capacity
Success rates	80% to 95% of BAP accepted for regeneration
Service life of regenerated BAP	Usage time can be multiplied by 2
Regenerative capacity	90% to 100% of manufacturer's capacity

3.3. COMPARATIVE ANALYSIS OF DIFFERENT REGENERATION PROCESSES

3.3.1. Technical analysis of regeneration processes

Table *3.5*: Technical comparison of processes

Technical data	Chemical process	Electrical process	Combined process
Battery types	Open (start-up, traction and stationary)	Open and sealed (start-up, traction and stationary)	Open and sealed (start-up, traction and stationary)
Use of the process in preventive maintenance	Yes	Yes	Yes
Process duration	Desulphation: a few hours Regeneration results: After several days	Desulphation: Three hours to four days depending on battery type and initial condition Regeneration results: After several charging and discharging cycles	Desulphation: Four hours to two days depending on battery type Results: After several weeks' use of the desulphated battery
Regeneration success rate	70 à 90 %	80 à 100 %	80 à 95 %

Criteria for verifying regeneration success	Regenerated battery capacity	Regenerated battery capacity	Regenerated battery capacity
Service life	Can be multiplied by 1.5	Can be multiplied by 1.75	Can be multiplied by 1.75
Regenerative capacity	85% to 95% of manufacturer's capacity	95% to 100% of manufacturer's capacity	90% to 100% of manufacturer's capacity

The three processes identified are capable of desulphating all types of lead-acid batteries: starter, traction and stationary. Within lead-acid batteries, it should be noted that chemical processes can only desulfate open batteries, i.e. those with a liquid electrolyte. As the chemical additive is usually in liquid (or soluble powder) form, it is technically easier to inject it into the battery if its electrolyte is also liquid.

The three processes are relatively similar in terms of preparation and finalization phases. The desulphation phases depend on the type of process used.

The desulphation steps themselves differ from process to process.

The chemical process involves injecting a chemical (in liquid, powder or capsule form) into the battery to desulfate it. The electrical process, on the other hand, involves applying short-duration currents to the battery terminals. The combined process combines these two methods. In this case, the chemical additive can be added before or after the electrical desulphation process, depending on the players involved.

3.3.2. **Environmental analysis of regeneration processes**

This analysis is based on the various flows linked to the different regeneration processes. The main flows linked to the implementation of the processes identified in this study are of four types:

- Incoming flows :
    - Energy consumption ;
    - Consumption of raw materials.
- Outgoing flows :
    - Waste generated ;

- Emissions to air and/or water.

Table 3.6 below shows the nature of incoming and outgoing flows for each process.

Table *3.6*: Environmental balance of processes

Related equipment	Chemical process	Electrical process	Combined process
		Incoming flows	
Energy	No electricity consumption directly linked to the process.	Depending on battery type and number of charge-discharge cycles	Depending on battery type and number of charge/discharge cycles (reduced compared to electric process)
Raw materials	- Chemical additives based on organic polymer, EDTA, hydrogen peroxide or other compounds, confidential compositions (33 ml per battery to 60 ml per 100 Ah) - Electrolyte for levelling	- Components and materials needed to manufacture desulphation units	- Chemical additives based on organic polymer, EDTA, hydrogen peroxide, phosphoric acid or other compounds, confidential compositions (33 ml per battery to 60 ml per 100 Ah) - Electrolyte for upgrade or replacement - Components and materials needed to manufacture desulphation units - Components and materials needed to manufacture electronic enclosures
		Outgoing flows	

Waste generated	- Batteries that cannot be desulphated - Waste associated with additive manufacture and storage (containers)	- Battery cells replaced - Batteries that cannot be desulphated	- Battery cells replaced - Batteries that cannot be desulphated - Waste associated with additive manufacture and storage (containers)
Emission into water and/or air	Information not available	Acid release	Hydrogen release

The main parameters influencing energy consumption are machine power and desulphation time.

Raw materials and their quantities are highly dependent on the type of process used. Processes with a chemical component require chemical raw materials that are not present in other types of process.

As for electrical processes, they use electrical and electronic components as raw materials, the exact composition of which is also unknown and dependent on the installation.

The main waste products from desulphation are battery cells or entire batteries, which were defective at the outset or which have not withstood desulphation.

In addition, for processes using a chemical additive, empty containers (cans) are waste. For players who manufacture their own additives, manufacturing-related waste (chemical compounds, containers for these compounds) can also be added to the desulphation environmental balance.

Emissions into the air (or water) linked to the desulphation process are essentially emissions linked to the operation and maintenance of lead batteries (acid, hydrogen), or emissions linked to the possible transport of batteries.

3.3.3. **Economic analysis of regeneration processes**

At this level, it is difficult to make an economic comparison of processes, as cost variations do not seem to be linked to processes, but rather to the organizational models of the players involved, and to the prices of the desulphation machines they use. As far as service provision is concerned, we can estimate the cost of setting up an average regeneration plant at between €25,000 and €100,000 (amount linked solely to the purchase of regeneration machinery and equipment) [13].

The price of regeneration services is the same for all players and processes: 30% to 60% of the price of a new battery for industrial batteries, 50% of the price of a new battery for starter batteries. [13].

With regard to process profitability, it has not been possible to obtain cost data from regeneration operators. Nevertheless, the development of desulphation services, whether networked or independent, seems to indicate that there is a market for this activity. The following chapter 4 will give us these profitability indicators, based on our technical choices, our organizational model, our assumptions and our study context.

CONCLUSION

This study has shown that the three processes identified are capable of desulfating and regenerating all types of BAP, i.e. start-up, traction and stationary.

The electrical process is capable of treating all types of open and sealed BAP without any complexity, while the chemical processes are most often used on open batteries. For sealed batteries, however, they consist in drilling a hole in the side of each cell at the usual electrolyte level, tapping it and then, after treatment, sealing it with a nylon screw.

The three processes are relatively similar in terms of preparation and completion phases. The desulphation phases for regeneration, on the other hand, depend on the type of process used, which will characterize the installation for the process concerned.

CHAPTER 4: DESIGNING A REGENERATION PROCESS FOR USED LEAD-ACID BATTERIES IN BENIN

INTRODUCTION

Following the technical study carried out, a technical choice of certain equipment characteristic of the installation and a study of the costs generated by the project to install a BAPU regeneration plant in Benin, as well as an environmental impact study, are necessary in order to evaluate the amounts and forecast risks that the installation of this plant could generate.

4.1. TECHNICAL CHOICES

The aim here is to specify and/or briefly argue the choice of the most important plant features related to the BAPU regeneration activity. These include the choice of :

- Unit location ;
- Regeneration process to be implemented ;
- Supplier of regeneration equipment ;
- Type of BAPU to be regenerated ;
- Regeneration service price ;

4.1.1. **Location**

A rural site will be chosen to be closer to solar installations, which are generally located in isolated areas.

4.1.2. **Regeneration process to be installed**

The regeneration process to be used is the electrical process, because of its efficiency, success rate, lower environmental impact, lower labor costs, especially since it can regenerate all types of batteries, including the most common ones encountered in our context, and the greater number of testimonials from service providers in the field compared with the chemical and combined processes.

4.1.3. **Supplier of regeneration equipment**

Based on the characteristics of the regenerators described in the literature and the regenerators available on the market, we chose the **BRT MAXI GOLD** regenerator, the most powerful regenerator in the **BATTERIE PLUS** regenerator range. What's more, its efficiency has been verified since 2005 [43] at the Laboratoire Central des Industries Electriques at the Bureau de Veritas in France.

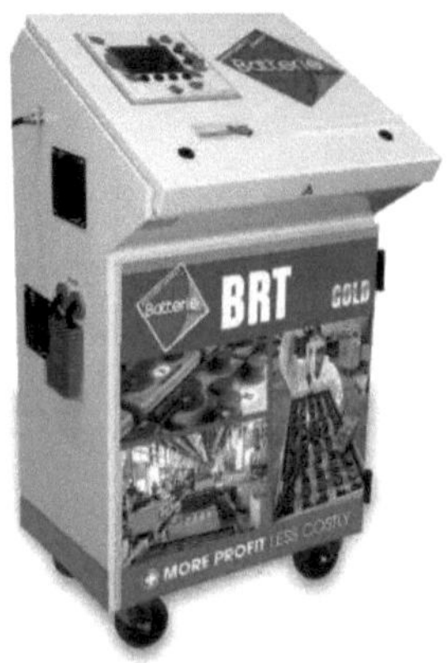

Figure 4.1: BRT MAXI GOLD regenerator

4.1.4. **Type of BAPU to be regenerated**

Because of their main uses in the solar installations listed above, the BAPUs to be regenerated are technically the following:

- **Gel** batteries **12 V / 150 Ah** ;
- **OPzV 2 V / 2000 Ah** batteries.

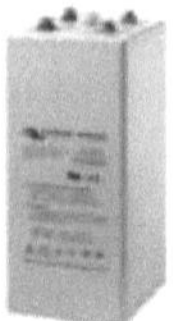

Figure *4.2*: Type of batteries involved

4.1.5. Regeneration service price

According to regeneration professionals, the price of the service varies between 30% and 60%. [13] of the price of a new battery. Our regeneration service will cost 35% of the price of a new battery of the same type, given the country's poverty-stricken situation.

4.2. FINANCIAL ANALYSIS

A number of assumptions and parameters were taken into account in this assessment. Evaluations were based on a 5-year operating period.

4.2.1 Business plan presentation

4.2.1.1. Objectives

- Set up a regeneration chain for the **20,000 BAPUs** quantified in Benin (**15,000 12 V / 150 Ah gels** and **5,000 2 V / 2,000 Ah OPzVs**);
- Supply the Beninese government with regenerated batteries at a flat rate of 35% of the price of a new battery;
- Supply **3,000 Gel 12 V / 150 Ah** and **1,000 OPzV 2 V / 2,000 Ah** in the first year and every 5 years thereafter.
- Adapt annual service capacity according to the proposed approach to process all 20,000 BAPUs over 5 years.

4.2.1.2. Main phases of the project

The main phases of the project are presented in Table 4.1 below.

Table 4.1: Main project phases

Phases	Duration
Administrative procedures	3 months
Ordering and receiving equipment	3 months
• System installation • Testing and commissioning • Technical staff training	1 month
Project follow-up (services, maintenance, equipment renewal)	15 years
Dismantling	7 Months + 15 years

4.2.1.3. General structure of initial investment

The initial working capital requirement for the project amounts to **148,456,504 F CFA (excl. VAT)**. The initial investment amounts to **336,452,504 F CFA (excl. VAT).** The general structure of the investment is shown in table 4.2 below.

Table 4.2: General structure of initial investment

Designation	Quantity	Cost per unit	Amount (F CFA)
Set-up costs			
Research costs; patent and trademark registration		8 000 000	
Advertising and launch costs		3 000 000	
Subtotal 1			-
Land			
Land	1	15 000 000	15 000 000
Construction	1	20 000 000	20 000 000
Head office	1	110 000 000	
Land title(fees)	1	2 000 000	
Notarial deeds	1	2 500 000	
Miscellaneous expenses	1	3 000 000	
Sub-total 2			**35 000 000**
Technical installations and fittings			
Complete workshop	1	93 796 000	93 796 000

Designation	Quantity	Cost per unit	Amount (F CFA)
Used forklift trucks	1	3 210 000	3 210 000
Miscellaneous ancillary equipment	1	7 350 000	7 350 000
Rackc/storage/bins	1	4 590 000	4 590 000
Stores	2	5 000 000	
Administration	1	5 000 000	
Interior courtyard	1	10 000 000	
Fire safety	1	2 500 000	2 500 000
Water drilling	1	3 000 000	
PV installation	1	250 000 000	
Subtotal 3			**111 446 000**
Software			-
Accounting, sales, payroll and asset management software	1	1 000 000	1 000 000
Subtotal 4			**1 000 000**
Technical equipment and tools			
Generator	1	-	
Subtotal 5			
Hardware			
Computers	10	400 000	4 000 000
Laser printer	2	275 000	550 000
Furniture	4	500 000	2 000 000
Photocopier	2	500 000	1 000 000
Subtotal 6			**7 550 000**
Transport equipment			
7 T delivery van	1	8 000 000	8 000 000
Liaison vehicle (4*4)	1	10 000 000	10 000 000
Subtotal 7			**18 000 000**
Working capital requirement (WCR)			148 456 504
Subtotal 8			**148 456 504**
Total investment			**336 452 504**

4.2.1.4. Financing strategy

The investment will be 34% equity and 66% external. Banks, organizations such as the African Development Bank (ADB), the World Bank or other investment banks, or even independent organizations and individuals wishing to contribute to this project, could be considered as external lenders. The financing strategy for this project is shown in table 4.3 below.

Table 4.3: Financing strategy

JOBS	AMOUNT	%	RESOURCES	AMOUNT	%
Fixed assets	187 996 000	56%	Shareholders' equity	114 050 000	34%
			Grant	0	0%
WCR	148 456 504	44%	Bank loan	222 402 504	66%
Total	**336 452 504**	**100%**		**336 452 504**	**100%**

4.2.1.5. Performance assumptions

This project will regenerate 20,000 BAPUs over 5 years. As the service is provided at 35% of the price of a new battery, the service price, service quantity and sales to be achieved each year are given in tables 4.4, 4.5 and 4.6 below.

Table 4.4: Prices for regeneration services

Designation / Batteries	Unit price of a new battery F CFA	Rates	Regeneration unit price FCFA
12 V / 150 Ah	175 000	0,35	61 250
2 V / 2000 Ah	720 500	0,35	252 175

Table 4.5: Quantity of services to be provided over 5 years

Designation	Year 1	Year 2	Year 3	Year 4	Year 5	Total

12 V / 150 Ah	3000	3000	3000	3000	3000	15000
2 V / 2000 Ah	1000	1000	1000	1000	1000	5000

Table 4.6: Sales estimates

Designation	Year 1	Year 2	Year 3	Year 4	Year 5
12 V / 200 Ah	183 750 000	183 750 000	183 750 000	183 750 000	183 750 000
2 V / 2000 Ah	252 175 000	252 175 000	252 175 000	252 175 000	252 175 000
Sales (FCFA)	**435 925 000**	**435 925 000**	**435 925 000**	**435 925 000**	**435 925 000**

4.2.1.6. Financial indicators of profitability

Using a discount rate of 12%, table 4.7 below presents the key figures and financial profitability indicators for our project.

The table shows that the project is highly profitable over a 5-year period. The 5-year profitability study yields the following financial indicators:

- A **Net Present Value (NPV) of CFAF 475,426,769;**
- A **profitability index close to 2.41** ;
- A **payback period of 1 year 11 months 11 days**;
- An **Internal Rate of Return (IRR) of 44%.**

Table 4.7 Financial indicators of project profitability

	AN 0	AN 1	AN 2	AN 3	AN 4	AN 5
NET INCOME		173 427 561	235 445 487	228 289 058	219 727 807	209 482 809
DEPRECIATION AND AMORTIZATION		28 199 200	64 747 509	62 779 491	60 425 147	57 607 773
CAF		**201 626 761**	**300 192 996**	**291 068 549**	**280 152 954**	**267 090 582**
VAR WCR	148 456 504	- 52 263 600	153 268 804	- 46 493 565	160 187 885	-
INVESTMENTS	187 996 000					
Grant						
Total capital expenditure	**336 452 504**	**- 52 263 600**	**153 268 804**	**- 46 493 565**	**160 187 885**	-
Net cash flow	- 336 452 504	253 890 361	146 924 192	337 562 114	119 965 069	267 090 582
Discount rate (12%)	1,000	0,893	0,797	0,712	0,636	0,567
FNT ACT	**- 336 452 504**	**226 687 823**	**117 127 066**	**240 270 045**	**76 239 970**	**151 554 369**
FNT ACT CUMULATED	**- 336 452 504**	**- 109 764 681**	**7 362 385**	**247 632 430**	**323 872 400**	**475 426 769**
VAN	**475 426 769**					
IP	**2,413057603**					
DRCI	**1 year 11 months 11 days**		706			
SHOOTING	**44%**					

4.2.2. Business plan summary tables

4.2.2.1.Investment and financing

These are the costs involved in purchasing equipment and supplies, and building infrastructure. Prices for processing equipment and machinery were provided by suppliers, while a lump sum was estimated for office supplies. As for the construction of the plant's infrastructure, as we did not have the necessary parameters to draw up a detailed construction estimate, the costs were estimated on the basis of reports from studies and projects in the building and civil engineering sector. The unit prices of the necessary components and the total estimated amounts were then derived. The results are shown in Table 4.8.

Table 4.8: Investment and financing

Designation	Quantity	Cost per unit	Amount (F CFA)
Set-up costs			
Research costs; patent and trademark registration		8 000 000	
Advertising and launch costs		3 000 000	
Subtotal 1			-
Land			
Land	1	15 000 000	15 000 000
Construction	1	20 000 000	20 000 000
Head office	1	110 000 000	
Land title(fees)	1	2 000 000	
Notarial deeds	1	2 500 000	
Miscellaneous expenses	1	3 000 000	
Sub-total 2			**35 000 000**
Technical installations and fittings			
Complete workshop	1	93 796 000	93 796 000
Used forklift trucks	1	3 210 000	3 210 000
Miscellaneous ancillary equipment	1	7 350 000	7 350 000
Rackc/storage/bins	1	4 590 000	4 590 000
Stores	2	5 000 000	
Administration	1	5 000 000	

Designation	Quantity	Cost per unit	Amount (F CFA)
Interior courtyard	1	10 000 000	
Fire safety	1	2 500 000	2 500 000
Water drilling	1	3 000 000	
PV installation	1	250 000 000	
Subtotal 3			**111 446 000**
Software			-
Accounting, sales, payroll and asset management software	1	1 000 000	1 000 000
Subtotal 4			**1 000 000**
Technical equipment and tools			
Generator	1	-	
Subtotal 5			
Hardware			
Computers	10	400 000	4 000 000
Laser printer	2	275 000	550 000
Furniture	4	500 000	2 000 000
Photocopier	2	500 000	1 000 000
Subtotal 6			**7 550 000**
Transport equipment			
7 T delivery van	1	8 000 000	8 000 000
Liaison vehicle (4*4)	1	10 000 000	10 000 000
Subtotal 7			**18 000 000**
Working capital requirement (WCR)			148 456 504
Subtotal 8			**148 456 504**
Total investment			**336 452 504**

4.2.2.2.Operating expenses

They take into account the purchase price of raw materials, personnel costs, taxes, maintenance, communication, insurance, water and electricity, and possibly the hire of service providers.

Personnel costs were calculated on the basis of the salaries received by each member according to their position, to which the percentage allocated to social charges was applied. In addition, a supplementary rate has been applied to employee salaries. This corresponds to a margin in relation to possible salary increases, bonuses and bonuses that may be awarded to employees. This rate has been set at an average of 23% of annual salary. Expenditure on consumables (water and electricity) was calculated on the basis of estimated consumption by suppliers according to the number of machines installed, and flat-rate costs for SONEB at 750 FCFA per m^3 incl. VAT, and for SBEE at 185 FCFA per KWh. The results are shown in table 4.9.

Table 4.9: Operating expenses

PURCHASES	AN1	AN 2	AN 3	AN 4	AN 5
Technical equipment	93 796 000	-	-	-	-
Supplies and consumables	6 960 000	7 140 000	7 338 000	7 555 800	7 795 380
External services	1 200 000	1 260 000	1 323 000	1 389 150	1 458 608
Telecommunications costs	1 200 000	1 260 000	1 323 000	1 389 150	1 458 608
Bank charges	198 240	208 152	218 560	229 488	240 962
Bank charges	198 240	208 152	218 560	229 488	240 962
Personnel expenses	49 003 200	58 803 840	70 564 608	84 677 530	101 613 036
Financial expenses	524 200	550 410	577 931	606 827	637 168
Depreciation and amortization	28 199 200	28 199 200	28 199 200	28 199 200	28 199 200
TOTAL	**172 920 840**	**89 021 602**	**100 883 298**	**115 102 194**	**132 148 973**

4.2.2.3.Depreciation

This parameter defines the loss in value of acquisitions made by the company: it enables the depreciation of an asset due to wear and tear, time or obsolescence to be recognized in the accounts. When an asset is purchased by a company, it is always assumed to be for a limited period of time. Once this period has elapsed, the company's accounting system considers the asset to be 100% depreciated, and its book value to be zero. The calculation of depreciation is shown in table 4.10.

Table 4.10: Calculating depreciation

Designation	Amount	Rates	Depreciation and amortization				
			Year 1	Year 2	Year 3	Year 4	Year 5
Set-up costs		33,33 %					
Construction	35 000 000	2,00%	700 000	700 000	700 000	700 000	700 000
Technical installations and fittings (PV)	111 446 000	20,00 %	22 289 200	22 289 200	22 289 200	22 289 200	22 289 200
Software	1 000 000	10,00 %	100 000	100 000	100 000	100 000	100 000
Hardware	7 550 000	20,00 %	1 510 000	1 510 000	1 510 000	1 510 000	1 510 000
Transport equipment	18 000 000	20,00 %	3 600 000	3 600 000	3 600 000	3 600 000	3 600 000
Total			**28 199 200**	**28 199 200**	**28 199 200**	**28 199 200**	**28 199 200**

4.2.2.4.Working capital requirement

It corresponds to the sum needed by the unit to pay its current expenses while waiting to receive the payment due from its customers. It's a factor that shows a company's financial autonomy. It corresponds to inventories and receivables minus payables to suppliers.

The calculation of WCR is based on assumptions reflecting a situation that could become critical for the company, in order to know how much it should plan for. The assumptions considered were that the company would have to cover the purchase of equipment and staff salaries for 6 months, knowing that the government owed it two months' payment and that it owed its suppliers 40% of the total cost of the equipment. The results are shown in table 4.11.

Table 4.11: Calculation of working capital requirements Forecast income statement

DESIGNATION	Hypothesis	AN 1	AN 2	AN 3	AN 4	AN 5
Needs to be financed						
Purchase of equipment		93 796 000	-	-	-	-
External services	6 months	600 000	630 000	661 500	694 575	729 304
Personnel expenses	6 months	19 920 000	23 904 000	28 684 800	34 421 760	41 306 112
Accounts receivable (days sales)	60	71 658 904	71 658 904	71 658 904	71 658 904	71 658 904
Total financing requirements		185 974 904	96 192 904	101 005 204	106 775 239	113 694 320
Resources						
Suppliers	40% of total equipment cost	37 518 400	-	-	-	-
Total resources		37 518 400	-	-	-	-
WCR		148 456 504	96 192 904	101 005 204	106 775 239	113 694 320
Change in WCR		**148 456 504**	**- 52 263 600**	**153 268 804**	**- 46 493 565**	**160 187 885**

4.2.2.5.Income statement

The income statement presents all of a company's income and expenses. By focusing on gains and losses, it provides a picture of net income (profit or loss). Operating income reflects the company's economic profitability. Based on these forecasts, it is already possible to expect to generate profits from the 2ème year of operation.

Table 4.12: Forecast income statement

DESIGNATION	AN 1	AN 2	AN 3	AN 4	AN 5
Sales figures	**435 925 000**	**435 925 000**	**435 925 000**	**435 925 000**	**435 925 000**
Purchases of raw materials	93 796 000	-	-	-	-
Gross margin on materials	**342 129 000**	**435 925 000**	**435 925 000**	**435 925 000**	**435 925 000**
Purchases of non-stockable supplies and consumables	6 960 000	7 140 000	7 338 000	7 555 800	7 795 380
External services	1 398 240	1 468 152	1 541 560	1 618 638	1 699 569
Taxes	198 240	208 152	218 560	229 488	240 962
Added value	**333 572 520**	**427 108 696**	**426 826 881**	**426 521 075**	**426 189 089**
Personnel expenses	39 840 000	47 808 000	57 369 600	68 843 520	82 612 224
Gross operating profit (EBITDA)	**293 732 520**	**379 300 696**	**369 457 281**	**357 677 555**	**343 576 865**
Depreciation and amortization	28 199 200	28 199 200	28 199 200	28 199 200	28 199 200
Operating income	**265 533 320**	**351 101 496**	**341 258 081**	**329 478 355**	**315 377 665**
Financial expenses	26 322 890	26 349 100	26 376 621	26 405 518	26 435 859
Profit before tax	**239 210 430**	**324 752 396**	**314 881 460**	**303 072 837**	**288 941 806**
Income tax	65 782 868	89 306 909	86 592 401	83 345 030	79 458 997
Net income	**173 427 561**	**235 445 487**	**228 289 058**	**219 727 807**	**209 482 809**
Depreciation and amortization	28 199 200	28 199 200	28 199 200	28 199 200	28 199 200
Financing capacity	**201 626 761**	**263 644 687**	**256 488 258**	**247 927 007**	**237 682 009**

4.2.2.6.Cash flow budget

The cash flow budget is used to analyze the company's activity from a purely banking point of view, in order to highlight the company's net cash position. It is made up of cash inflows and outflows. Inflows are the sums that the company is expected to receive over the forecast period, in this case sales. Disbursements are the sums that the company will pay out over the forecast period, i.e. expenses and annual repayments to external investors.

Table 4.13: Cash flow budget

DESIGNATION	AN 1	AN 2	AN 3	AN 4	AN 5
Receipts					
Sales incl. VAT	514 391 500	514 391 500	514 391 500	514 391 500	514 391 500
Shareholders' equity	114 050 000				
Bank loan	222 402 504				
Total receipts	**850 844 004**	**514 391 500**	**514 391 500**	**514 391 500**	**514 391 500**
Disbursement					
Purchasing materials	110 679 280	-	-	-	-
Overheads	8 358 240	8 608 152	8 879 560	9 174 438	9 494 949
Investments	187 996 000				
Taxes + financial expenses	722 440	758 562	796 490	836 315	878 130
Personnel expenses	49 003 200	58 803 840	70 564 608	84 677 530	101 613 036
Refund		48 038 941	48 038 941	48 038 941	48 038 941
Total disbursements	**356 759 160**	**116 209 495**	**128 279 599**	**142 727 223**	**160 025 056**
CASH BALANCE	**494 084 844**	**398 182 005**	**386 111 901**	**371 664 277**	**354 366 444**
FINAL CASH POSITION	**494 084 844**	**892 266 849**	**1 278 378 751**	**1 650 043 028**	**2 004 409 472**

4.2.2.7.Rate of return

These rates are: overall business profitability, economic profitability and financial profitability.

The first value measures the company's overall capacity, the second measures the company's capacity to generate profits from capital invested, and the third measures the company's capacity to remunerate associates:

- The first rate corresponds to the ratio of net operating income to sales;
- The second rate corresponds to the ratio of operating income to invested capital (fixed assets + WCR);
- The third rate corresponds to the ratio of operating income to shareholders' equity.

Table 4.14: Rates of return

DESIGNATION	AN 1	AN 2	AN 3	AN 4	AN 5
Overall profitability (Income/Sales)	39,78 %	54,01%	52,37%	50,40%	48,05%
Return on investment (ROI)	82,60 %	109,22 %	106,16 %	102,50 %	98,11%
Net return on equity (ROE)	152%	206%	200%	193%	184%

4.2.2.8.Financial profitability indicators : Investment selection criteria

Table 4.15: Financial profitability indicators: Investment selection criteria

	AN 0	AN 1	AN 2	AN 3	AN 4	AN 5
NET INCOME		173 427 561	235 445 487	228 289 058	219 727 807	209 482 809
DEPRECIATION AND AMORTIZATIO N		28 199 200	64 747 509	62 779 491	60 425 147	57 607 773
CAF		**201 626 761**	**300 192 996**	**291 068 549**	**280 152 954**	**267 090 582**
VAR WCR	148 456 504	- 52 263 600	153 268 804	- 46 493 565	160 187 885	-

INVESTMENTS	187 996 000					
Grant						
Total capital expenditure	**336 452 504**	**- 52 263 600**	**153 268 804**	**- 46 493 565**	**160 187 885**	-
Net cash flow	- 336 452 504	253 890 361	146 924 192	337 562 114	119 965 069	267 090 582
Discount rate (12%)	1,000	0,893	0,797	0,712	0,636	0,567
FNT ACT	**- 336 452 504**	**226 687 823**	**117 127 066**	**240 270 045**	**76 239 970**	**151 554 369**
FNT ACT CUMULATED	**- 336 452 504**	**- 109 764 681**	**7 362 385**	**247 632 430**	**323 872 400**	**475 426 769**
VAN	**475 426 769**					
IP	**2,413057603**					
DRCI	**1 year 11 months 11 days**		706			
SHOOTING	**44%**					

4.3. ENVIRONMENTAL ASSESSMENT OF THE USED LEAD BATTERY REGENERATION ACTIVITY

Basically, it has not been possible to obtain information on the negative impacts associated with BAPU regeneration from professionals in the field, or from previous studies that have looked into this aspect of regeneration, considering it to be a green activity. We therefore thought it would be useful to give an overview of the environmental impact of this activity in our context. This involves citing the possible impacts and risks from installation to operation of the business.

Since the exact location of the site is not yet known, it can be said for the moment that there may or may not be any direct effects on fauna, flora or the human environment, although it should be noted that there are potential risks associated with the operation of BAP.

4.3.1. Potential risks when handling lead batteries

Handling BAP presents the following potential risks:

- Sulfuric acid in the electrolyte can cause severe burns;
- Lead is a toxic substance;
- The hydrogen and oxygen released can be explosive;
- High electrical currents may be generated, and electric shocks may occur in the event of short circuits.

All these risks are already taken into account on the machines, making the electric process even simpler, faster, more efficient and cleaner than other BAPU regeneration processes.

CONCLUSION

In order to determine whether our business is financially feasible, we undertook a study of the costs involved in setting it up. The main parameters evaluated were investments, expenses, depreciation, WCR, income statement, sales, cash flow budget and profitability indicators. Based on these results, it would be fair to say that the BAPU regeneration plant is profitable from its 3ème year of operation. In addition to being financially profitable, such a project will bring a great deal to Benin in terms of job creation, particularly green jobs, which are very well supported by sustainable development organizations and players. Indeed, not only will the company employ many people, but it will also have an indirect impact on the job market through the partnerships it could form.

GENERAL CONCLUSION

This study was undertaken with the aim of gaining an insight into the quantity of BAPU, the best regeneration process and the financial profitability of setting up a BAPU regeneration plant in Benin.

For the BAPU evaluation, we based ourselves on the achievements of the Beninese government through several projects (PRODERE, PROVES, etc...) in terms of streetlight and solar micropower installations. With over 20,000 streetlights and some four dozen solar micropower plants already installed, and other projects in progress, the current potential is estimated at **19,132 Gel 12 V / 150 Ah BAPUs** and **6,262 OPzV 2 V / 2,000 Ah BAPUs.**

After a technical and comparative study, **the electrical regeneration process** was found to be more economical, efficient, less waste-generating and more environmentally friendly than chemical and combined processes. This will enable the installation of a virtually automatic, and therefore faster, system. Nevertheless, it would be preferable for further studies to shed more light on the environmental characterization of all three regeneration processes mentioned in this study.

Once these first two steps had been taken, it proved important to carry out a financial analysis of the project in order to determine its profitability. This showed that, with an operating life of **5 years**, the total investment cost was **475,426,769 FCFA, and the** payback period was **1 year 11 months 11 days.**

This financial study was carried out with the aim of regenerating batteries throughout Benin, but also with a view to extending the BAPU regeneration activity to the West African sub-region in the long term, where only a few exist at present.

It should be noted that battery regeneration is above all :

- A cost-effective alternative for end-users, who pay 50 to 60% less than for a new battery;
- Unrivalled responsiveness, with immediate service and no need to wait for delivery;

- The fight against programmed obsolescence ;
- Massive reduction of hazardous and polluting industrial waste
- The creation of new jobs as part of a virtuous circular economy;
- Carbon footprint reduction 50 times less carbon-intensive than recycling and replacement with new batteries;
- An alternative for reducing the operating costs of photovoltaic systems, making them more competitive;
- Increasing the country's territorial resilience.

Benin imports almost all of its electricity from neighboring Nigeria, but with its drive towards energy transition, and in view of the country's solar installation programs, the need for back-up for industrial, institutional and administrative users in the event of grid outages, and the increasing number of off-grid solar installations given the country's considerable solar potential, the battery regeneration market has a bright future ahead of it.

So, with such potential, wouldn'tit be more economical to install a plant equipped with a BAPU regeneration workshop and a BAPU recycling plant to recycle non-regenerable batteries?

BIBLIOGRAPHY

[1] " Optimizing the life of lead batteries - Lesson V02-Bis.pdf ".

[2] H. H. Molinaro and B. Multon, "Technologies for electrical energy storage systems", p. 21.

[3] F. Sanchez, "Etat de l'art sur le recyclage et le réemploi des batteries", p. 176.

[4] E. Korsaga, Z. Koalaga, D. Bonkougou, and F. Zougmoré, "Comparison and determination of appropriate storage devices for a stand-alone photovoltaic system in a Sahelian zone," 2018, doi: 10.18145/JITIPEE.V4I1.161.

[5] C. Glaize, "À CHAQUE APPLICATION, UNE TECHNOLOGIE DE BATTERIE ADAPTÉE", p. 43, 2014.

[6] " IEEE Standard Glossary of Stationary Battery Terminology," IEEE. doi: 10.1109/IEEESTD.2016.7552407.

[7] G. L. Soloveichik, "Battery Technologies for Large-Scale Stationary Energy Storage," *Annu. Rev. Chem. Biomol. Eng.* vol. 2, n° 1, pp. 503-527, Jul. 2011, doi: 10.1146/annurev-chembioeng-061010-114116.

[8] World Health Organization, *Recycling used lead-acid batteries: health considerations*. Geneva: World Health Organization, 2017. Accessed: 24 December 2021. [Online]. Available at: https://apps.who.int/iris/handle/10665/259446

[9] C. Consulting and K. Environmental, *Environmentally sound management of used lead-acid batteries in North America.* Montreal: Commission for Environmental Cooperation, 2016. Accessed: December 24, 2021. [Online]. Available at: http://public.ebookcentral.proquest.com/choice/publicfullrecord.aspx?p=4532547

[10] " Study of the effect of acid density in the formation of PbO2.pdf ".

[11] W. Jamratnaw, "Desulfation of lead-acid battery by high frequency pulse," in *2017 14th International Conference on Electrical Engineering/Electronics, Computer, Telecommunications and Information Technology (ECTI-CON)*, Phuket, June 2017, pp. 676-679. doi: 10.1109/ECTICon.2017.8096328.

[12] A. Thierry, "and real-world simulation of autonomous solar systems based on the KEG method", p. 97.

[13] " State of the art of desulfation technologies for lead-acid batteries", p. 76, 2011.

[14] " How a solar battery works _ SOLARIS-STORE.html ".

[15] " EHS-DOC-146_LeadAcidBatteries.pdf ".

[16] "treatment-de-regeneration-desulfatation-des-batteries-electronique-chimique-edta-sulfate-de-.pdf".

[17] " GFA06_EN_Financial analysis and project evaluation.pdf ".

[18] F. Delahaye-Duprat and J. Delahaye, *Finance d'entreprise: manuel*, 7th ed [Levallois-Perret] Malakoff: Éditions Francis Lefebvre Dunod, 2018.

[19] D. SEWANOUDE, "Manuel de cours de l'Analyse Economique et Financière des Projets". GME/EPAC, 2019.

[20] B. Lucie, "Elaboration d'une méthode de calcul de retour sur investissement applicable aux projets immobiliers des établissements publics de santé. Réflexions à partir des projets du Centre Hospitalier Universitaire de Nancy", p. 105, 2008.

[21] G. J. May, A. Davidson, and B. Monahov, "Lead batteries for utility energy storage: A review," *J. Energy Storage*, vol. 15, pp. 145-157, Feb. 2018, doi: 10.1016/j.est.2017.11.008.

[22] S. J. Hou, Y. Onishi, S. Minami, H. Ikeda, M. Sugawara, and A. Kozawa, "Charging and Discharging Method of Lead Acid Batteries Based on Internal Voltage Control," *J. Asian Electr. Veh.* vol. 3, n° 1, pp. 733-737, 2005, doi: 10.4130/jaev.3.733.

[23] C. Beaurain and M. B. Amoussou, "Les enjeux du développement de l'énergie solaire au Bénin. Quelques pistes de réflexion pour une approche territoriale:", *Mondes En Dév.*, vol. n° 176, n° 4, p. 59-76, Dec. 2016, doi: 10.3917/med.176.0059.

[24] " Rapport_situation_éclairage_public_Final_ABERME_2020.pdf ".

[25] K. G. M. Yêhouénou, "Transition énergétique au Bénin: quel apport du solaire photovoltaïque?", p. 292.

[26] " RAPPORT_DE_SYNTHESE_étude_ESEIM_2019.pdf ".

[27] " REPORT EF MICROCENTRALE DEF.pdf ".

[28] J. Schiffer, D. U. Sauer, H. Bindner, T. Cronin, P. Lundsager, and R. Kaiser, "Model prediction for ranking lead-acid batteries according to expected lifetime in renewable energy systems and autonomous power-supply systems", *J. Power Sources*, vol. 168, n° 1, pp. 66-78, May 2007, doi: 10.1016/j.jpowsour.2006.11.092.

[29] H. Ikeda, S. Minami, H. Wada, and A. Kozawa, "Outline of Three Main Businesses of Lead-acid Batteries Using ITE's Organise Polymer Activators", *J. Asian Electr. Veh.* vol. 5, n° 1, pp. 987-988, 2007, doi: 10.4130/jaev.5.987.

[30] S. Ikeda, "Innovations of Lead-Acid Batteries", *Electrochemistry*, vol. 76, n° 1, pp. 32-37, 2008, doi: 10.5796/electrochemistry.76.32.

[31] "minami2004.pdf".

[32] A. Paglietti, "Electrolyte Additive Concentration for Maximum Energy Storage in Lead-Acid Batteries," *Batteries*, vol. 2, n° 4, p. 36, Nov. 2016, doi: 10.3390/batteries2040036.

[33] L. T. Lam *et al*, "Pulsed-current charging of lead/acid batteries - a possible means for overcoming premature capacity loss?", *J. Power Sources*, vol. 53, n° 2, pp. 215-228, Feb. 1995, doi: 10.1016/0378-7753(94)01988-8.

[34] "2_541.pdf".

[35] Yan Zhang, Songjie Hou, Shigeyuki Minami, and Akiya Kozawa, "A high current pulse activator for the prolongation of Lead-acid batteries", in *2008 IEEE Vehicle Power and Propulsion Conference*, Harbin, Hei Longjiang, China, Sept. 2008, p. 1-4. doi: 10.1109/VPPC.2008.4677701.

[36] " HAYDEN, Matthew, P. et al; Taft, Stettinius &.pdf ".

[37] S. Oumar, "26500 BOURG LES VALENCE (FR)", p. 26.

[38] N. S. M. Ibrahim *et al*, "Parameters observation of restoration capacity of industrial lead acid battery using high current pulses", *Int. J. Power Electron.*

Drive Syst. IJPEDS, vol. 11, n° 3, p. 1596, Sept. 2020, doi: 10.11591/ijpeds.v11.i3.pp1596-1602.

[39] A. C. Ohajianya, E. C. Mbamala, C. M. Amakom, and C. E. Akujor, "An empirical investigation of lead-acid battery desulfation using a high-frequency pulse desulfator", *J. Adv. Sci. Eng.* vol. 4, n° 1, pp. 44-52, Jan. 2021, doi: 10.37121/jase.v4i1.140.

[40] " Afrique Avenir: regenerating lead-acid batteries, Marly Diallo's environmental boost - BBC News Afrique". https://www.bbc.com/afrique/56665416 (accessed March 3, 2022).

[41] " UBC (Univers Business CHAD) leader de la régénération des batteries au Tchad - Battery Regeneration". https://batteryregeneration.net/ubc-univers-business-chad-leader-de-la-regeneration-des-batteries-au-tchad/ (accessed March 3, 2022).

[42] H. Ikeda, S. Minami, S. J. Hou, Y. Onishi, and A. Kozawa, "Nobel High Current Pulse Charging Method for Prolongation of Lead-acid Batteries," *J. Asian Electr. Veh.* vol. 3, n° 1, pp. 681-687, 2005, doi: 10.4130/jaev.3.681.

[43] "reportLCIE.pdf".

TABLE OF CONTENTS

SUMMARY 2

GENERAL INTRODUCTION 3

CHAPTER 1: GENERAL INFORMATION ON BATTERIES 5

INTRODUCTION 5

1.1 BATTERY OPERATING PRINCIPLE 6

1.2 DIFFERENT TYPES OF ELECTROCHEMICAL BATTERIES 7

1.2.1. Battery types by field of application 7

1.2.2. Battery types by technology 9

1.3. GENERAL INFORMATION ON LEAD ACID TECHNOLOGY 11

1.3.1. Composition of a lead-acid battery 11

1.3.2. Operating principle of a lead-acid battery 13

1.3.3. State of discharge and charge of a lead-acid battery 13

1.3.4. Main features 15

1.3.5. Typology of lead-acid batteries 17

1.4. RECOVERY PROCESSES FOR LEAD-ACID BATTERIES 20

1.4.1. Recycling 20

1.4.2. Regeneration 21

1.5. TYPE OF FINANCIAL ANALYSIS 22

1.5.1. Criteria for assessing the financial profitability of a project 22

CONCLUSION 25

CHAPTER 2: DEGRADATION OF LEAD-ACID BATTERIES AND POTENTIAL OF USED LEAD-ACID BATTERIES IN BENIN 26

INTRODUCTION 26

2.1. CAUSES OF DEGRADATION 26

2.1.1. Corrosion of the positive grid 26

2.1.2. Positive network growth 26
2.1.3. Sulfation 27
2.1.4. Softening of the active ingredient 28
2.1.5. Acidic stratification 28
2.1.6. Drying 28
2.1.7. Abutment joint leakage 29
2.1.8. Cover gasket leak 29
2.1.9. Vent failure 29
2.1.10. Mechanical damage 29
2.1.11. Corrosion of the group bar 29
2.1.12. Internal short circuit 29
2.1.13. External hydrogen ignition 30
2.1.14. Overheating of external connections 30
2.1.15. Thermal runaway 30

2.2. MOST FREQUENT MODES OF DEGRADATION 30

2.2.1. Corrosion of the positive electrode 31
2.2.2. Degradation of the active ingredient 32
2.2.3. Sulfation 32

2.3. ASSESSING THE POTENTIAL OF BAPU IN BENIN 33

2.3.1. Current status of solar street lighting in Benin 35
2.3.2. Current status of PV micropower plants in Benin 38

CONCLUSION 41

CHAPTER 3: TECHNICAL STUDY OF LEAD BATTERY REGENERATION 43

INTRODUCTION 43

3.1. GENERAL INFORMATION ON REGENERATION 43

3.1.1. Why desulphation? 43
3.1.2. The benefits of regeneration 43
3.1.3. Definition 44
3.1.4. Regeneration principle 44

3.2. STATE-OF-THE-ART BATTERY REGENERATION PROCESSES 45

3.2.1. Chemical regeneration process 45
3.2.2. Electrical regeneration process 49

3.2.3. Combined regeneration process 54

3.3. COMPARATIVE ANALYSIS OF DIFFERENT REGENERATION PROCESSES 57

3.3.1. Technical analysis of regeneration processes 57

3.3.2. Environmental analysis of regeneration processes 58

3.3.3. Economic analysis of regeneration processes 61

CONCLUSION 61

CHAPTER 4: DESIGNING A REGENERATION PROCESS FOR USED LEAD-ACID BATTERIES IN BENIN 62

INTRODUCTION 62

4.1. TECHNICAL CHOICES 62

4.1.1. Location 62

4.1.2. Regeneration process to be installed 62

4.1.3. Supplier of regeneration equipment 63

4.1.4. Type of BAPU to be regenerated 63

4.1.5. Price of regeneration services 64

4.2. FINANCIAL ANALYSIS 64

4.2.1 Presentation of the business plan 64

4.2.2. Business plan summary tables 70

4.3. ENVIRONMENTAL ASSESSMENT OF THE USED LEAD BATTERY REGENERATION ACTIVITY 78

4.3.1. Potential risks when handling lead batteries 78

CONCLUSION 80

GENERAL CONCLUSION 81

BIBLIOGRAPHY 83

TABLE OF CONTENTS 87

Printed by Books on Demand GmbH, Norderstedt / Germany